Sets
Relations
and
Groups

IBH : Topic 8

Steven Clarke
United Nations International School
New York

This book covers the material required for the Option Topic (Sets, Relations and Groups) in the new syllabus of IB Higher Level Mathematics.

The book can be used for self-study, as numerous worked examples are provided .

If you find any typographical errors, or incorrect answers, please let me know at   sclarke@unis.org

Also available   **VECTORS   IBH : TOPIC 4**
ISBN 978-1478393566

SECOND EDITION

Copyright © 2017   Steven Clarke

Previous edition :   Copyright © 2015   Steven Clarke

ISBN 978-1535177221

Contents

Presumed Knowledge :

Functions
Venn Diagrams
Trigonometry

4

Notation:

$\{x_1, x_2, x_3, \ldots\ldots, x_n\}$ a set containing n elements

$\in$ is an element of

$\notin$ is not an element of

$|$ such that

$:$ such that

$n(A)$ number of elements in set A

U   universal set

$\varnothing$ empty set

A'   complement of set A

$\subseteq$ is a subset of

$\subset$ is a proper subset of

$[a, b]$ $a \leq x \leq b$

$]a, b[$ $a < x < b$

$\Leftrightarrow$ if and only if

A\B   set difference : $A - B = A \cap B'$

$A \triangle B$ symmetric difference of two sets : A\B $\cup$ B\A

gcd (a, b) the greatest common divisor of integers a and b

lcm (a, b) the least common multiple of integers a and b

Sets

A set is a collection of similar items (known as *elements*).

A set is always well-defined.
(In other words, you always know if a particular element belongs to a given set).

Example :   Given $A = \{3, 4, 5, 6, 7, 8\}$ we know $5 \in A$ and $2 \notin A$

  (Note : $5 \in A$ reads   "5 is an element of set A"
     $2 \notin A$ reads   "2 is not an element of set A")

Example :   Given $B = \{x \mid 7 \leq x \leq 10, \; x \in \mathbb{N} \}$
     we know that $8 \in B$ and $5 \notin B$

  (Note : $\{x \mid 7 \leq x \leq 10, \; x \in \mathbb{N} \}$ reads   "the set of numbers x
     such that $7 \leq x \leq 10$ and x is a natural number ".
     Hence $B = \{7, 8, 9, 10\}$)

In a set , the order of the elements is not important.
$\{1, 2, 3, 4\}$ is the same as $\{4, 2, 1, 3\}$

Also, in a set, elements are not repeated.
$\{a, a, b\}$ is the same as $\{a, b\}$

$n(A)$ = number of elements in set A
If $A = \{3, 4, 5, 6, 7, 8\}$ then $n(A) = 6$
If $B = \{x \mid 7 \leq x \leq 10, \; x \in \mathbb{N} \}$ then $n(B) = 4$

A and B are called **finite** sets.

Common sets :

$\mathbb{N}$ = the set of natural numbers $\{0, 1, 2, 3, 4, \ldots\ldots\}$

$\mathbb{Z}$ = the set of integers $\{\ldots\ldots -3, -2, -1, 0, 1, 2, 3, \ldots\ldots\}$

$\mathbb{Z}^{+}$ = the set of positive integers $\{1, 2, 3, \ldots\ldots\}$

$\mathbb{Z}^{-}$ = the set of negative integers $\{-1, -2, -3, \ldots\ldots\}$

$\mathbb{Q}$ = the set of rational numbers

$\mathbb{R}$ = the set of real numbers

$\mathbb{C}$ = the set of complex numbers

$\mathbb{R} - \{0\}$ = the set of real numbers except for zero.

$\mathbb{Z}^{+} = \mathbb{N} - \{0\}$

$\mathbb{R} - \mathbb{Q}$ = the set of irrational numbers (numbers which cannot be expressed in the form $\dfrac{a}{b}$ where $a, b \in \mathbb{Z}$)

$\mathbb{C} - \mathbb{R}$ = $\{x+iy \mid i = \sqrt{-1}, \ y \neq 0 \text{ and } x, y \in \mathbb{R}\}$

U = universal set (the set of everything)

(Note: "everything" will be defined for a given situation)

$\varnothing$ = empty set (the set that contains no elements)

$= \{\ \}$

Infinite Sets

$\mathbb{N} = \{0, 1, 2, 3, 4, \ldots\ldots\}$ is an **infinite** set.

$A = \{x \mid 7 \leq x \leq 10, x \in \mathbb{R}\}$ is also an infinite set

because there are an infinite number of real numbers between 7 and 10.

Union $\cup$

Given two sets: $A = \{3, 4, 5, 6, 7, 8\}$ and $B = \{7, 8, 9, 10\}$

The union of the two sets is $\{3, 4, 5, 6, 7, 8, 9, 10\}$

This can be written as $A \cup B$

It is the set of elements which are either in A **or** in B **or** in both A and B.

This can be represented on a Venn Diagram by the following shaded area :

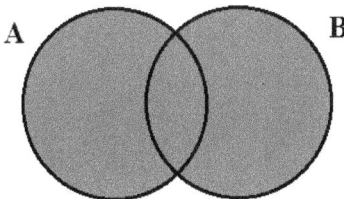

Note: $A \cup \varnothing = A$

$A \cup U = U$

Intersection ∩

Given two sets: A = {3, 4, 5, 6, 7, 8} and B = {7, 8, 9, 10}

The intersection of the two sets is {7, 8}

This can be written as A ∩ B

It is the set of elements which are in both A **and** B.

This can be represented on a Venn Diagram by the following shaded area :

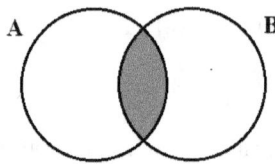

Note: A ∩ Ø = Ø

   A ∩ U = A

Complement

Suppose the universal set U = {1, 2, 3, 4, 5, 6, 7, 8, 9, 10}
and A = {3, 4, 5, 6, 7, 8} .

The complement of set A is {1, 2, 9, 10}

This can be written as A'

These are all of the elements not in set A (but contained in U)

Hence   A ∪ A' = U

   A ∩ A' = Ø

Set difference

$$A - B = A\backslash B$$

$$= A \cap B'$$

$$= \text{elements in set A that are not in set B}$$

This can be represented on a Venn Diagram by the following shaded area :

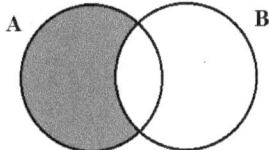

Symmetric difference Δ

$$A \, \Delta \, B = (A - B) \cup (B - A)$$

$$= (A \cap B') \cup (B \cap A')$$

$$= (A \cup B) - (A \cap B)$$

This can be represented on a Venn Diagram by the following shaded area :

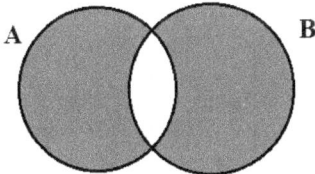

Subsets

If set A is contained in set B then A is a subset of B.

This is written as $A \subseteq B$

Formal definition : If $x \in A \Rightarrow x \in B$ then $A \subseteq B$

Given the set $A = \{ a, b, c, d \}$ the possible subsets of A are

$\{ a, b, c, d \}$
$\{ a, b, c \}$
$\{ a, b, d \}$
$\{ a, c, d \}$
$\{ b, c, d \}$
$\{ a, b \}$
$\{ a, c \}$
$\{ a, d \}$
$\{ b, c \}$
$\{ b, d \}$
$\{ c, d \}$
$\{ a \}$
$\{ b \}$
$\{ c \}$
$\{ d \}$
$\varnothing$

The set of all possible subsets of A is called the 'power set' of A, written as P(A).

If $\{a, b, c, d\}$ is excluded then the subsets are called **proper** subsets.

If a set contains n elements then there will always be 2^n possible subsets.
(and $2^n - 1$ proper subsets)

In the example above, set A has 4 elements and hence 16 possible subsets.

Proving two sets are equal

To prove that set A = set B it is necessary to show that $A \subseteq B$ **and** $B \subseteq A$

(This is similar to the following algebraic implication :
$x \le y$ and $y \le x \implies x = y$)

Example: If $A \cup B = A \cap B$ prove that $A = B$

Proof

Let $x \in A$
It then follows that $x \in A \cup B$ since any element in A must also be in $A \cup B$.
Since we are told that $A \cup B = A \cap B$, we now know that $x \in A \cap B$.
Hence $x \in B$, because any element in $A \cap B$ must also be in B.

We have now shown that $x \in A \implies x \in B$
which means that A is a subset of B

Let $x \in B$
Then $x \in A \cup B$ since any element in B must also be in $A \cup B$
Since $A \cup B = A \cap B$, we now know that $x \in A \cap B$
Hence $x \in A$ because any element in $A \cap B$ must also be in A

We have now shown that $x \in B \implies x \in A$
which means that B is a subset of A

Taken together $A \subseteq B$ **and** $B \subseteq A \implies A = B$

Note : In the IBH exam this type of proof is only used for simple expressions.
It would not be used to prove , for example , that

$(A \cap B') \cup (A' \cap B) = (A \cup B) \cap (A \cap B)'$

Here you would use set algebra for the proof (see later)

DeMorgans Rule

$$(A \cup B)' = A' \cap B'$$

$$(A \cap B)' = A' \cup B'$$

This rule is easily illustrated on a Venn Diagram :

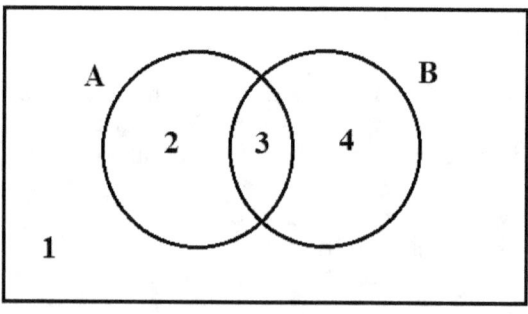

$(A \cup B)'$ is represented by area 1

    A' is represented by areas 1 and 4

    B' is represented by areas 1 and 2

    $A' \cap B'$ is the overlap of A' and B' which is also area 1

$(A \cap B)'$ is represented by areas 1 , 2 and 4

    A' is represented by areas 1 and 4

    B' is represented by areas 1 and 2

    $A' \cup B'$ is the union of A' and B' which is also areas 1, 2, and 4

A Venn Diagram is not considered a " proof " in IBH math.
The method shown on Page 13 should be used.

Example: Prove $(A \cup B)' = A' \cap B'$ (DeMorgans Rule)

Proof

Let $x \in (A \cup B)'$

Then $x \notin A \cup B$

Hence $x \notin A$ and $x \notin B$ since x cannot be in either set

This gives $x \in A'$ and $x \in B'$

Hence $x \in A' \cap B'$

Which gives $(A \cup B)' \subseteq A' \cap B'$

Let $x \in A' \cap B'$

Then $x \in A'$ and $x \in B'$

Hence $x \notin A$ and $x \notin B$

This gives $x \notin A \cup B$ since x cannot be in either set

Hence $x \in (A \cup B)'$

Which gives $A' \cap B' \subseteq (A \cup B)'$

Since $(A \cup B)' \subseteq A' \cap B'$ and $A' \cap B' \subseteq (A \cup B)'$

We have $(A \cup B)' = A' \cap B'$

Set Algebra

$$A \cup (B \cap C) = (A \cup B) \cap (A \cup C)$$

$$A \cap (B \cup C) = (A \cap B) \cup (A \cap C)$$

This corresponds to the distributive law : $x(y + z) = xy + xy$

Example: Simplify $A \cap (B \cup A')$

$$
\begin{aligned}
A \cap (B \cup A') &= (A \cap B) \cup (A \cap A') \\
&= (A \cap B) \cup \varnothing \\
&= A \cap B
\end{aligned}
$$

Example: Simplify $A \cup (A \cup B')'$

$$
\begin{aligned}
A \cup (A \cup B')' &= A \cup (A' \cap B) \text{ by De Morgans Rule} \\
&= (A \cup A') \cap (A \cup B) \\
&= U \cap (A \cup B) \\
&= A \cup B
\end{aligned}
$$

16

Example: Simplify $(A \cap B) \cup (A \cap B')$

$$(A \cap B) \cup (A \cap B') = A \cap (B \cup B')$$
$$= A \cap U$$
$$= A$$

Note : compare this to $xy + xz = x(y + z)$

Expanding

$$(A \cap B) \cup (C \cap D) = (A \cup C) \cap (A \cup D) \cap (B \cup C) \cap (B \cup D)$$

$$(A \cup B) \cap (C \cup D) = (A \cap C) \cup (A \cap D) \cup (B \cap C) \cup (B \cap D)$$

This corresponds to $(x + y)(a + b) = xa + xb + ya + yb$

Example: Show that $(A \cap B') \cup (A' \cap B) = (A \cup B) \cap (A \cap B)'$

$$(A \cap B') \cup (A' \cap B) = (A \cup A') \cap (A \cup B) \cap (B' \cup A') \cap (B' \cup B)$$
$$= U \cap (A \cup B) \cap (B' \cup A') \cap U$$
$$= (A \cup B) \cap (B' \cup A')$$
$$= (A \cup B) \cap (A \cap B)'$$

Cartesian Product

$$A \times B = \{ (x, y) \mid x \in A \text{ and } y \in B \}$$

Example: If $A = \{1, 2, 3\}$ and $B = \{a, b\}$

then $A \times B = \{ (1, a), (1, b), (2, a), (2, b), (3, a), (3, b) \}$

and $B \times A = \{ (a, 1), (a, 2), (a, 3), (b, 1), (b, 2), (b, 3) \}$

$\mathbb{R} \times \mathbb{R}$ = the set of coordinate points in the Cartesian plane.

$\mathbb{R} \times \mathbb{R}$ is often written as $\mathbb{R}^2$

Exercise 1

1) $U = \{x \mid 1 \le x \le 10, \ x \in \mathbb{N}\}$ and A and B are subsets of U.

$A = \{x \mid x \text{ is a prime number}\}$

$B = \{1, 2, 3, 5, 8, 9\}$

Write down   (i) $A \cap B$

(ii) $(A \cup B)'$

(iii) $A' \cup B$

(iv) $A - B$

(v) $A \vartriangle B$

2) U = {positive integers less than 20} and P and Q are subsets of U.

$P = \{x \mid x^2 - 12x + 20 \le 0\}$

$Q = \{x \mid x^2 \in U\}$

Find   (i)   the elements of set P

(ii)   the elements of set Q

(iii) $P \cap Q$

(iv) $P \backslash Q$

(v) $P' \cap Q'$

(vi) $P \vartriangle Q$

3) Given $n(A \cup B) = 20$, $n(A) = 15$, $n(B) = 12$, calculate $n(A \cap B)$

4) Simplify (i) $\mathbb{Z}^+ \cup \mathbb{Z}^- \cup \{0\}$

            (ii) $\mathbb{Z} - \mathbb{N}$

            (iii) $\varnothing'$

            (iv) $n(\varnothing)$

            (v) $n(\{\varnothing\})$

5) Using the properties of sets prove that $A \cap (A \cap B')' = A \cap B$ justifying each step of the proof.

6) Using set properties, prove that $(A \cap B)' \cap (A \cup B) = A \triangle B$ justifying each step of the proof.

7) Use a Venn Diagram to show that $(A \cap B) \cup (A \cap B') = A$

8) If $A \cap B = A$ prove that $A \subseteq B$

9) If $A = \{1, 2, 3\}$ and $B = \{4, 5\}$ list the elements of

     (i) $A \times B$

     (ii) $B \times A$

10) Using set properties, prove that $(A \cap B) - (A \cap C) = A \cap (B - C)$

Relations

A relation connects two sets.
The two sets are called domain and range.

aRb means element a relates to element b under some condition.
For example, $a = 2b$, or $a < b$, or "a has the same nationality as b"

A relation produces a set of ordered pairs.

Example :

Given the relation $R = \{ (1,6)\ (2,6)\ (2,12)\ (4, 7)\ (4, 9)\ (5,6)\ (5, 8) \}$

The domain of R is $\{1, 2, 4, 5\}$. These are the x-coordinates.
The range of R is $\{6, 7, 8, 9, 12\}$. These are the y-coordinates.

[Note : (a, b) indicates aRb]

Example:

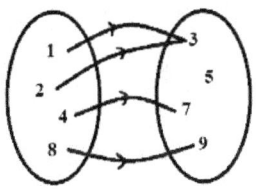

The domain of this relation is $\{1,2,4,8\}$

The range of the relation is $\{3,7,9\}$

The set $\{3, 5, 7, 9\}$ is called the co-domain.

[Note : the range is a subset of the co-domain.]

Functions

A function is a relation in which every element of the domain maps onto one, and only one, element in the range.

The relation $R = \{(1, 3), (2, 3), (4, 7)\}$ is a function with domain $\{1, 2, 4\}$ and range $\{3, 7\}$

The relation $R = \{(1, 3), (2, 4), (2, 5)\}$ is **not** a function because element 2 in the domain maps onto both elements 4 **and** 5 in the range.

$f(x) = 2x + 1$, $x \in \mathbb{Z}^+$ is a function with domain $\{1, 2, 3, \ldots\ldots\}$
and range $\{3, 5, 7, \ldots\ldots\}$

$f(x) = x^2$, $x \in \mathbb{R}$ is a function with domain $\mathbb{R}$ and range $\mathbb{R}^+ \cup \{0\}$

[Note: Domain = {all possible x values}
     Range = {all possible y values}]

It is often helpful to sketch a graph of the function in order to determine the range.

$f : \mathbb{R} \to \mathbb{R}$ indicates domain $\mathbb{R}$ and co-domain $\mathbb{R}$

$f : \mathbb{Z} \to \mathbb{N}$ indicates domain $\mathbb{Z}$ and co-domain $\mathbb{N}$

Inverse

The inverse of a relation maps each element of the range back onto the corresponding element in the domain.

Example: The inverse of $R = \{ (1, 3), (2, 4), (2, 5) \}$ is given by
$R^{-1} = \{ (3, 1), (4, 2), (5, 2) \}$

[Note: although R is not a function, R^{-1} is a function.]

Example: The inverse of $f(x) = 2x + 1$, $x \in \mathbb{Z}^+$ is $f^{-1}(x) = \dfrac{x - 1}{2}$

This can be found by one of two methods :

1) Observation : the function $f(x)$ takes any element in the domain, multiplies it by 2 and adds 1. Working backwards, $f^{-1}(x)$ will take any element in the range, subtract 1 and divide by 2.

Hence we obtain $\dfrac{x - 1}{2}$

2) Let $y = 2x + 1$.
   To find the inverse, switch x and y to give $x = 2y + 1$
   Re-arrange to make y the subject : $y = \dfrac{x - 1}{2}$

Note : the domain of $f^{-1}(x)$ is $\{3, 5, 7, 9, \ldots\ldots\}$

Simple rule: the domain of $f^{-1}(x)$ is the range of $f(x)$

Example: Find the inverse of $f(x) = \dfrac{3x-1}{x+2}$, $x \in \mathbb{R} - \{-2\}$

Method: Let $y = \dfrac{3x-1}{x+2}$

For inverse, switch x and y : $\quad x = \dfrac{3y-1}{y+2}$

Re-arrange to make y the subject :

$$x(y+2) = 3y - 1$$

$$xy + 2x = 3y - 1$$

$$2x + 1 = 3y - xy$$

$$2x + 1 = y(3 - x)$$

$$\frac{2x+1}{(3-x)} = y$$

Hence $f^{-1}(x) = \dfrac{2x+1}{3-x}$, $\quad x \in \mathbb{R} - \{3\}$

Example: Find the inverse of $g(x) = 4^x + 2^x + 1$, $x \in \mathbb{R}$

Method: Let $y = 4^x + 2^x + 1$

For inverse, switch x and y: $x = 4^y + 2^y + 1$

To make y the subject is not easy !
You need to use the quadratic formula

$$x = (2^2)^y + 2^y + 1$$

$$x = 2^{2y} + 2^y + 1$$

$$x = (2^y)^2 + 2^y + 1$$

$$(2^y)^2 + 2^y + 1 - x = 0$$

$$2^y = \frac{-1 \pm \sqrt{1 - 4(1-x)}}{2}$$

$$2^y = \frac{-1 \pm \sqrt{4x-3}}{2}$$

$$y = \log_2 \frac{-1 + \sqrt{4x-3}}{2}$$

Note the change from $\pm$ to $+$
(it is not possible to have the log of a negative number.)

Hence $g^{-1}(x) = \log_2 \dfrac{-1 + \sqrt{4x-3}}{2}$, $x > 1$

[Note: we need $-1 + \sqrt{4x-3} > 0$. Hence domain $x > 1$]

The graphs of $g(x)$ and $g^{-1}(x)$ are shown below :

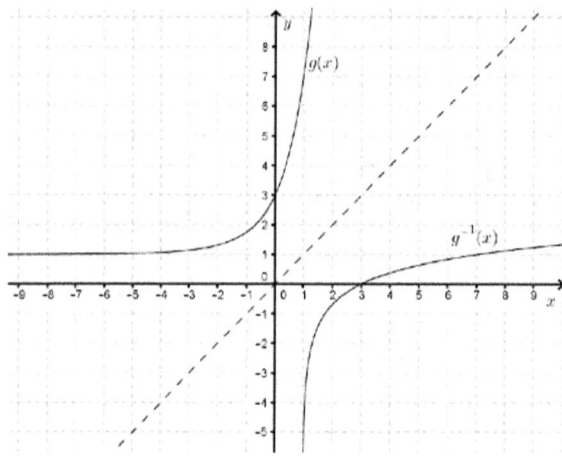

The domain of $g^{-1}(x)$ is easily seen from the graph.
It is the same as the range of $g(x)$.

Vertical line test

Any vertical line drawn onto the graph of a function will cut the graph only once.

Example:

The graph of a cubic polynomial represents a function

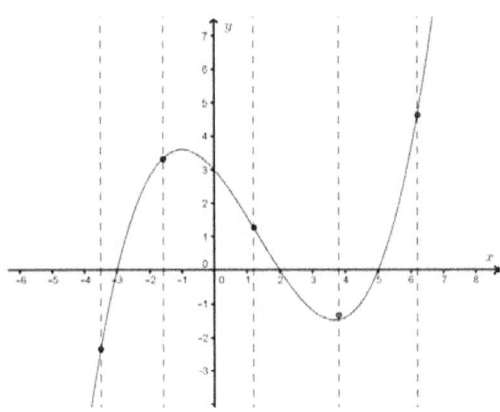

The graph of $y = \pm\sqrt{x}$ is not a function

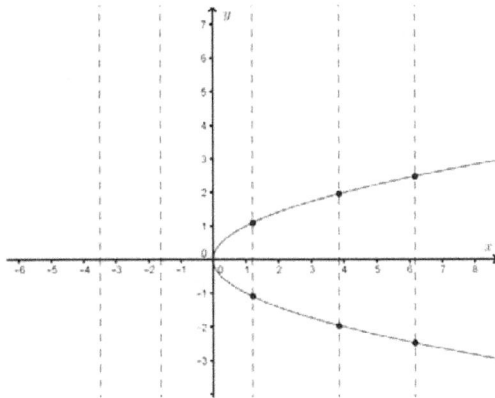

Injection

A function is an injection if every element in the co-domain is mapped onto by *at most* one element from the domain.

In other words , if $f(a) = f(b)$ then $a = b$

Example: Show that $f : \mathbb{R} \to \mathbb{R}$, $f(x) = 2x + 1$ is an injection.

$$f(a) = f(b) \quad \text{gives} \quad 2a + 1 = 2b + 1$$
$$a = b$$

Hence injection.

Example: Show that $f : \mathbb{R} \to \mathbb{R}$, $f(x) = \sin x$ is not an injection.

$f(a) = f(b)$ gives $\sin a = \sin b$
which does not imply $a = b$ since
$$\sin \frac{\pi}{6} = \sin \frac{5\pi}{6} \quad \text{and} \quad \frac{\pi}{6} \neq \frac{5\pi}{6}$$

Hence not an injection.

Example: Show that $f : \mathbb{R} \to \mathbb{R}$, $f(x) = x^2$ is not an injection.

$f(a) = f(b)$ gives $a^2 = b^2$ which does not imply $a = b$
since $a = 2$ and $b = -2$ is a possible solution.

Hence not an injection.

In graphical terms : a function is an injection if every horizontal line drawn through the graph of the function cuts the graph *at most* once.

Example: $f(x) = 2x + 1$, $x \in \mathbb{R}$ is an injection

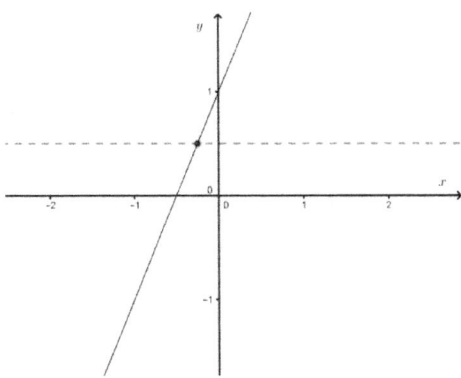

Example: $f(x) = \sin x$, $x \in \mathbb{R}$ is not an injection

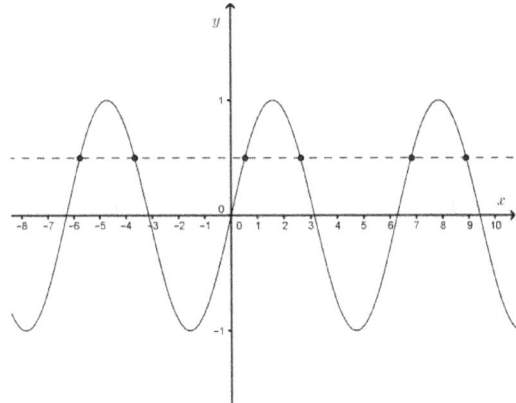

However, $f(x) = \sin x$, $x \in [-\dfrac{\pi}{2}, \dfrac{\pi}{2}]$ is an injection.

By restricting the domain to an appropriate size , any function can be made to be injective.

An injective function is also called a **one-to-one** function.

Surjection

A function is a surjection if **every** element in the co-domain has a corresponding element in the domain.

In other words , the range and co-domain are identical.

Example: Show that $f : \mathbb{R} \to \mathbb{R}$, $f(x) = 2x + 1$ is a surjection.

Every element n in the co-domain corresponds to $\dfrac{n-1}{2}$ in the domain. Hence it is a surjection.

Example: Show that $f : \mathbb{R} \to \mathbb{R}$, $f(x) = \sin x$ is not a surjection.

Every element outside $[-1, \ 1]$ in the co-domain does **not** have a corresponding element in the domain.
For example , $\sin x = 2$ has no solution
(the element 2 in the co-domain does **not** have a corresponding element in the domain).

In graphical terms : a function is a surjection if every horizontal line drawn through the graph of the function cuts the graph **at least** once.
(Note: it is essential that you take into consideration the domain and co-domain that is stated in the question).

A surjective function is also called an "**onto**" function.

Note: By restricting the co-domain so that it matches the range of the function we can produce a surjection.
Hence $f : \mathbb{R} \to [-1, \ 1]$, $f(x) = \sin x$ is a surjection.

Bijection

A function is a bijection if it is both an injection **and** a surjection.

A bijective function is also called a **one-to-one correspondence**.

Summary

not injective and not surjective	not injective but surjective
injective but not surjective	injective and surjective

The following graphs are all $f : \mathbb{R} \to \mathbb{R}$

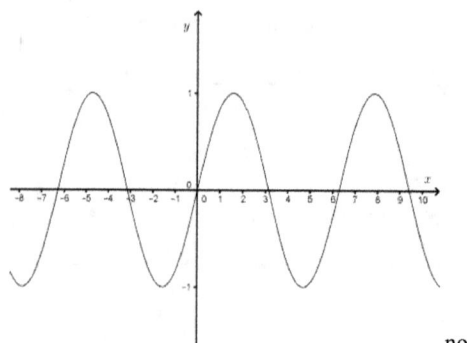

not injective and not surjective

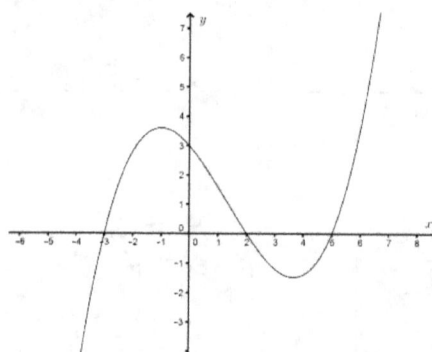

not injective but surjective

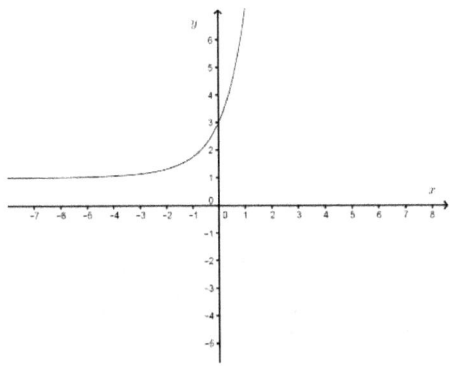

injective but not surjective

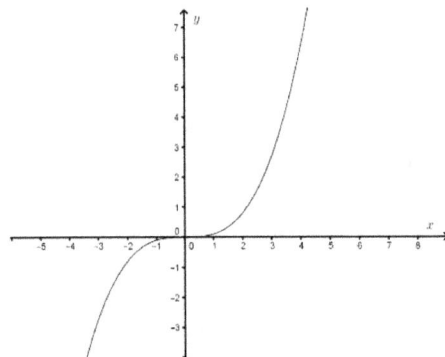

injective and surjective
(i.e. bijective)

Functions involving ordered pairs

This type of function maps one ordered pair onto another ordered pair ; for example $(4,3) \rightarrow (7,1)$

Here the domain and range will consist of sets of coordinate points.

Typical question :

Given $f: \mathbb{R} \times \mathbb{R} \rightarrow \mathbb{R} \times \mathbb{R}$ where $f(x,y) = (x+y , x-y)$

(a) calculate $f(6,2)$

(b) calculate (a, b) such that $f(a,b) = (14, 4)$

(c) show that $f(x,y)$ is injective

(d) find $f^{-1}(x,y)$

(e) show that $f(x,y)$ is surjective

(f) show that $f(x,y)$ is a bijection

Answer :

(a) $f(6,2) = (8, 4)$

(b) $f(a,b) = (a+b , a-b)$
$= (14, 4)$

$\Rightarrow a+b = 14$ and $a-b = 4$

$\Rightarrow a = 9 , b = 5$

(c)   For injection we need $f(a,b) = f(c,d) \Rightarrow (a,b)=(c,d)$

Here $f(a,b) = f(c,d) \Rightarrow (a+b,\ a-b) = (c+d,\ c-d)$

$$\Rightarrow a+b = c+d \quad \text{and}$$
$$a-b = c-d$$

adding gives $2a = 2c$ or $a = c$
subtracting gives $2b = 2d$ or $b = d$

Hence $(a,b)=(c,d)$

(d)   Let $f(a,b) = (p,q)$
Then $(a+b,\ a-b) = (p,q)$

$$a+b = p$$
$$a-b = q$$

adding gives $2a = p+q$

$$a = \frac{p+q}{2}$$

subtracting gives $2b = p-q$

$$b = \frac{p-q}{2}$$

Hence $f^{-1}(x,y) = (\frac{x+y}{2},\ \frac{x-y}{2})$

(e)   Every element (m,n) in the co-domain corresponds to element
$(\frac{m+n}{2},\ \frac{m-n}{2})$ in the domain.

Hence $f(x,y)$ is surjective.

(f)   Since $f(x,y)$ is injective and surjective it must be bijective.

Example: Find the inverse of $f: \mathbb{R}^2 \to \mathbb{R}^2$ where $f(x, y) = (2x - y , 3x + y)$

Let $f(a, b) = (p, q)$

Then $(2a - b , 3a + b) = (p, q)$

$$2a - b = p$$
$$3a + b = q$$

adding gives $5a = p + q$

$$a = \frac{p + q}{5}$$

Re-write the two equations as
$$6a - 3b = 3p$$
$$6a + 2b = 2q$$

subtracting gives $-5b = 3p - 2q$

$$b = \frac{3p - 2q}{-5}$$

$$= \frac{2q - 3p}{5}$$

Hence $f^{-1}(x, y) = (\frac{x + y}{5} , \frac{2y - 3x}{5})$

Exercise 2

1) State the range of the following functions :

    (a) $R = \{ (5, 3)\ (6, 3)\ (4, 8)\ (7, 8)\ (2, 9) \}$

    (b) $f(x) = e^x$, $x \in \mathbb{R}$

    (c) $f(x) = e^x$, $x \in \mathbb{R}^+$

    (d) $g(x) = x^2 - 10x + 24$, $x \in \mathbb{R}$

    (e) $h(x) = \dfrac{3x+1}{x+1}$, $x \in \mathbb{R} - \{-1\}$

2) Find the inverse of the following functions :

    (a) $f(x) = \dfrac{2x+1}{x-3}$, $x \in \mathbb{R} - \{3\}$

    (b) $f(x) = \sqrt{2x+8}$, $x \in [-4, \infty[$

    (c) $f(x) = x^2 - 3x + 2$, $x \in \mathbb{R}$

    (d) $f(x) = e^{2x} + e^x + 1$, $x \in \mathbb{R}$

3)   Given $f : \mathbb{R} \to \mathbb{R}$, $f(x) = e^x$

        (a) Prove (i) $f(x)$ is an injection

                (ii) $f(x)$ is not a surjection

        (b) Write down the inverse of $f(x)$

4) Given $f : \mathbb{R} \to \mathbb{R}$, $f(x) = x^3$ prove that $f(x)$ is a bijection.

5) Given that $f(x) = x^2 - 10x + 24$, $x \in \mathbb{R}$ is surjective, calculate the required co-domain.

6) Given $f : \mathbb{R}^2 \to \mathbb{R}^2$ where $f(x, y) = (2x + 3y , 4x - y)$

   (a) calculate $f(1, 2)$

   (b) calculate (a, b) such that $f(a, b) = (12, 10)$

   (c) find $f^{-1}(x, y)$

   (d) show that $f(x, y)$ is a bijection

Equivalence relations

If aRb satisfies the following three conditions then aRb is said to be an equivalence relation :

      1) aRb is reflexive

      2) aRb is symmetric

      3) aRb is transitive

Reflexive :   elements relate to themselves
                i.e. aRa

Symmetric :   if a relates to b then b relates to a
                i.e. $aRb \Rightarrow bRa$

Transitive: if a relates to b, and b relates to c, then a relates to c
                i.e. aRb and $bRc \Rightarrow aRc$

Example 1

      M = {males alive in the USA}
      R = the relation "born in the same year"

      Does R form an equivalence relation on set M ?

Yes, because

1) a person is born in the same year as himself.

2) if *Alex* is born in the same year as *Boris* then *Boris* is born in the same year as *Alex*.

3) if *Alex* is born in the same year as *Boris*, and *Boris* is born in the same year as *Chris*, then *Alex* is born in the same year as *Chris*.

Example 2

L = {straight lines}
R = the relation "is parallel to"

Does R form an equivalence relation on set L ?

Yes, because

1) a line is parallel to itself.

2) if line a is parallel to line b then line b is parallel to line a

3) if line a is parallel to line b, and line b is parallel to line c, then line a is parallel to line c.

Example 3

L = {straight lines}
R = the relation "is perpendicular to"

Does R form an equivalence relation on set L ?

No, because a line is not perpendicular to itself. Hence R is not reflexive.

It is also not transitive. If a is perpendicular to b, and b is perpendicular to c, then a is **parallel** (and not perpendicular) to c.

The relation is , however , symmetric because if a is perpendicular to b then b is perpendicular to a.

Equivalence Classes

An equivalence relation forms a **partition** on the set.
In other words, it divides the set into sections that each have something in common.
These sections are called **equivalence classes**.

In Example 1 , the equivalence classes are "birth years"

Born in 1898	......	Born in 1982	Born in 1983	Born in 1984	......	Born in 2017

The partition assumes that the oldest male person alive in the USA was born in 1898 and the current year is 2017.
Here the partition is finite – there are 120 equivalence classes.

In Example 2 , the equivalence classes are "lines with the same gradient".
Here the partition is infinite.

Definition

The equivalence class containing a is the set of elements that relate to a
i.e. $\{x : xRa\}$.

In other words, all of the elements that have something in common with a.

Example 4

Given aRb if and only if $a^3 \equiv b^3 \pmod 5$

prove that R is an equivalence relation on the set $\mathbb{N}$.

State the equivalence classes.

Note: $x \equiv y \pmod 5$ is defined by $x - y = 5m$ where $m \in \mathbb{Z}$.

In simple terms, y is the remainder when x is divided by 5.

For example, $18 = 3 \pmod 5$ and $27 = 2 \pmod 5$.

Proof

To prove that R is an equivalence relation we need to show that R is reflexive, symmetric and transitive.

We have $aRb \iff a^3 \equiv b^3 \pmod 5$

or $aRb \iff a^3 - b^3 = 5m$ where $m \in \mathbb{Z}$

aRa is valid because $a^3 - a^3 = 5m$ is true (for $m = 0$)

Hence R is reflexive.

$aRb \iff a^3 - b^3 = 5m$
$\Rightarrow b^3 - a^3 = -5m$
$\Rightarrow b^3 - a^3 = 5n$ where $n \in \mathbb{Z}$ (and $n = -m$)
$\Rightarrow bRa$

Hence R is symmetric.

$aRb \iff a^3 - b^3 = 5m$ and $bRc \iff b^3 - c^3 = 5n$
$\Rightarrow a^3 - b^3 + b^3 - c^3 = 5m + 5n$
$\Rightarrow a^3 - c^3 = 5(m + n)$
$\Rightarrow a^3 - c^3 = 5p$ where $p \in \mathbb{Z}$ (and $p = m + n$)
$\Rightarrow aRc$

Hence R is transitive.

To calculate the equivalence classes we need to look at each element of the set $\mathbb{N}$ under the relation R and find a pattern.

$$0^3 = 0 \ (\text{mod}\,5) \qquad\qquad 8^3 = 2 \ (\text{mod}\,5)$$
$$1^3 = 1 \ (\text{mod}\,5) \qquad\qquad 9^3 = 4 \ (\text{mod}\,5)$$
$$2^3 = 3 \ (\text{mod}\,5) \qquad\qquad 10^3 = 0 \ (\text{mod}\,5)$$
$$3^3 = 2 \ (\text{mod}\,5) \qquad\qquad 11^3 = 1 \ (\text{mod}\,5)$$
$$4^3 = 4 \ (\text{mod}\,5) \qquad\qquad 12^3 = 3 \ (\text{mod}\,5)$$
$$5^3 = 0 \ (\text{mod}\,5) \qquad\qquad 13^3 = 2 \ (\text{mod}\,5)$$
$$6^3 = 1 \ (\text{mod}\,5) \qquad\qquad 14^3 = 4 \ (\text{mod}\,5)$$
$$7^3 = 3 \ (\text{mod}\,5) \qquad\qquad 15^3 = 0 \ (\text{mod}\,5)$$

It can be seen that the elements $\{\,1, 6, 11, 16, 21, \ldots\ldots\,\}$ all result in $1 \ (\text{mod}\,5)$.

Similarly $\{2, 7, 12, 17, 22, \ldots\ldots\}$ all result in $3 \ (\text{mod}\,5)$.

$\{3, 8, 13, 18, 23, \ldots..\}$ all result in $2 \ (\text{mod}\,5)$.

$\{4, 9, 14, 19, 24, \ldots..\}$ all result in $4 \ (\text{mod}\,5)$.

$\{0, 5, 10, 15, 20, 25, \ldots..\}$ all result in $0 \ (\text{mod}\,5)$.

These are the 5 equivalence classes.

Hence the set $\mathbb{N}$ is partitioned into 5 sections by the relation R :

0, 5, 10, ...	1, 6, 11, ...	2, 7, 12, ...	3, 8, 13, ...	4, 9, 14, ...

Example 5

Question :

(a) Which elements must be added to the following set in order to make R an equivalence relation ?

$$R = \{ (5, 3) \ (3, 3) \ (4, 2) \ (2, 4) \ (3, 1) \ (4, 4) \ (2, 2) \}$$

[Note: (5, 3) indicates that $5R3$]

(b)   Find the equivalence classes.

Answer:

(a) We need R to be reflexive so we must have $1R1$, $2R2$, $3R3$, $4R4$ and $5R5$
Hence we need to add (1, 1) and (5, 5)

We also need R to be symmetric so $5R3$ requires $3R5$
Similarly $3R1$ requires $1R3$
Hence we need to add (3, 5) and (1, 3)

Finally, we need R to be transitive so $5R3$ and $3R1$ require $5R1$
So we need to add (5, 1) , and also (1, 5) to maintain symmetry.

(b) To find the equivalence classes we need to look for a pattern:

$\{ (5, 3) \ (3, 5) \ (3, 3) \ (5, 5) \ (3, 1) \ (1, 3) \ (1, 1) \ (5, 1) \ (1, 5) \}$ all inter-connect.

$\{ (4, 2) \ (2, 4) \ (4, 4) \ (2, 2) \}$ are separate.

These are the two equivalence classes.

Example 6

Question

(a) Given $(a,b)R(c,d) \Leftrightarrow ad = bc$ show that R forms an equivalence relation on $\mathbb{R} \times \mathbb{R}$.

(b) Find the equivalence class that contains $(1, 2)$

(c) Describe the partition that R creates on $\mathbb{R} \times \mathbb{R}$

Answer

(a) $(a,b)R(a,b) \Rightarrow ab = ba$ which is true. Hence R is reflexive.

$$
\begin{aligned}
(a,b)R(c,d) &\Rightarrow ad = bc \\
&\Rightarrow bc = ad \\
&\Rightarrow cb = da \\
&\Rightarrow (c,d)R(a,b) \qquad \text{Hence } R \text{ is symmetric.}
\end{aligned}
$$

$$
\begin{aligned}
(a,b)R(c,d) \text{ and } (c,d)R(e,f) \quad &\Rightarrow \quad ad = bc \text{ and } cf = de \\
&\Rightarrow \quad adcf = bcde \\
&\Rightarrow \quad af = be \\
&\Rightarrow \quad (a,b)R(e,f) \quad \text{Hence } R \text{ is transitive.}
\end{aligned}
$$

Since R is reflexive, symmetric and transitive it is an equivalence relation.

(b) To find the required equivalence class, consider a general point (x, y) in $\mathbb{R} \times \mathbb{R}$ together with $(1, 2)$

$$(x, y)R(1, 2) \Leftrightarrow 2x = y$$

This represents a straight line through the origin with slope 2. Hence the equivalence class is the set of coordinate points that lie on the line $y = 2x$.

(c) R partitions $\mathbb{R} \times \mathbb{R}$ into an infinite number of equivalence classes. Each class consists of coordinate points that lie on a straight line through the origin.

Exercise 3

1) R is a relation on the set of triangles T such that aRb if and only if triangle a is similar to triangle b.

    (a) Show that R is an equivalence relation on T

    (b) Describe the equivalence classes formed by R

2) (a) Which elements must be added to the following set in order to make R an equivalence relation ?

$$R = \{ (1, 2)\ (2, 3)\ (3, 3)\ (4, 4)\ (4, 5) \}$$

    (b) Write down the two equivalence classes.

3) Given $aRb \Leftrightarrow a^2 \equiv b^2 \pmod 2$

    (a) Prove that R is an equivalence relation on the set $\mathbb{N}$.

    (b) State the two equivalence classes.

4) Given $aRb \Leftrightarrow ab \geq 0$ explain why R is **not** an equivalence relation on the set $\mathbb{Z}$.

5) Given $aRb \Leftrightarrow a + b = 2n$ where $a, b, n \in \mathbb{Z}$.

    (a) Prove that R is an equivalence relation on the set $\mathbb{Z}$.

    (b) State the two equivalence classes of R.

Binary Operations

A binary operation combines two elements of a set to produce a third element.

$$a * b = c \quad \text{(where * is the binary operation)}$$

Examples : $3 + 4 = 7$
$5 \times 4 = 20$
$12 - 5 = 7$

Closure

Let $a, b \in M$

A binary operation $*$ is **closed** on a set M if $a * b \in M$

Example : The binary operation $+$ is closed on the set $\mathbb{N}$ because
the addition of two natural numbers always produces a
third natural number.

The binary operation $-$ is **not** closed on the set $\mathbb{N}$ because
the subtraction of two natural numbers could produce a negative
number. For example $6 - 8 = -2$ and $-2 \notin \mathbb{N}$

Identity

A binary operation $*$ on a set M has an **identity element** e if

$$a * e = e * a = a \qquad \text{for all } a \in M$$

Examples : $5 + 0 = 0 + 5 = 5$ (0 is the identity element for addition)

$3 \times 1 = 1 \times 3 = 3$ (1 is the identity element for multiplication)

Inverse

A binary operation $*$ on a set M has an **inverse** a^{-1} if

$$a*a^{-1} = a^{-1}*a = e \qquad \text{for all } a \in M$$

Example :   Every element a in set $\mathbb{Z}$ has an inverse $-a$ under the binary operation $+$

$$a + (-a) = (-a) + a = 0$$

$$4 + (-4) = (-4) + 4 = 0$$

Every element a in set $\mathbb{Q} - \{0\}$ has an inverse $\dfrac{1}{a}$ under the binary operation $\times$

$$a \times \frac{1}{a} = \frac{1}{a} \times a = 1$$

$$4 \times \frac{1}{4} = \frac{1}{4} \times 4 = 1$$

Associativity

The binary operation $*$ is said to be **associative** on a set M if

$$(a * b) * c = a * (b * c) \qquad \text{for all elements in the set}$$

Example: $(4 + 5) + 6 = 4 + (5 + 6)$ so addition is associative

$(16 \div 8) \div 2 \neq 16 \div (8 \div 2)$ so division is not associative

Commutativity

The binary operation $*$ is said to be **commutative** on a set M if

$a * b = b * a$ for all elements in the set.

Example : $4 + 5 = 5 + 4$ so addition is commutative.

$16 \div 8 \ne 8 \div 16$ so division is not commutative.

Note :   Binary operations are often defined by an expression.

For example $a * b = ab + a + b$

$a \, \Delta \, b = a^2 - b$

Hence, in this case $4 * 2 = 8 + 4 + 2$
$= 14$

$5 \, \Delta \, 3 = 25 - 3$
$= 22$

Example

Question:

The binary operation $*$ is defined by $a * b = ab + a + b$ for all $a, b \in \mathbb{R}$

    (a) (i) Show that $(3 * 2) * 4 = 3 * (2 * 4)$

          (ii) Prove that $*$ is associative

    (b) Find the identity element for $*$

    (c) (i) Find the inverse of $a \in \mathbb{R}$, stating any restrictions.

          (ii) Hence find the inverse of 5

Answer:

    (a) (i)

$$(3 * 2) * 4 = (6 + 3 + 2) * 4$$

$$= 11 * 4$$

$$= 44 + 11 + 4$$

$$= 59$$

$$3 * (2 * 4) = 3 * (8 + 2 + 4)$$

$$= 3 * 14$$

$$= 42 + 3 + 14$$

$$= 59$$

(ii) To prove $*$ is associative we need to show that $(a * b) * c = a * (b * c)$

$$(a * b) * c = (ab + a + b) * c$$
$$= (ab + a + b) c + (ab + a + b) + c$$
$$= abc + ac + bc + ab + a + b + c$$

$$a * (b * c) = a * (bc + b + c)$$
$$= a(bc + b + c) + a + (bc + b + c)$$
$$= abc + ab + ac + a + bc + b + c$$
$$= abc + ac + bc + ab + a + b + c$$

Hence $*$ is associative

(b) For identity we need to find a unique value of e such that

$$a * e = e * a = a \qquad \text{for all } a \in \mathbb{R}$$
$$a * e = a \quad \text{gives} \quad ae + a + e = a$$
$$ae + e = 0$$
$$e (a + 1) = 0$$
$$e = 0$$

$e * a = a$ gives the same result (this should always be checked)

(c) (i) For inverse we need to find a^{-1} such that $a*a^{-1} = a^{-1}*a = e$

Here $e = 0$

$$a*a^{-1} = 0 \quad \text{gives} \quad a\,a^{-1} + a + a^{-1} = 0$$

$$a^{-1}(a+1) = -a$$

$$a^{-1} = \frac{-a}{a+1}$$

$a^{-1}*a = 0$ gives the same result (again, this should always be checked)

The restriction is that $a \neq -1$

(ii) The inverse of 5 is $\dfrac{-5}{6}$

Example

Question:

The binary operation $*$ is defined by $a * b = a + b - 4$ for all $a, b \in \mathbb{N}$

    (a) Show that $*$ is not closed

    (b) Find the identity element for $*$

    (c) Find the inverse of $a \in \mathbb{N}$, stating any restrictions.

Answer:

    (a) To show that $*$ is not closed it is sufficient to give one example.
Since $a, b \in \mathbb{N}$ we just need to find any $a * b \notin \mathbb{N}$

$$1 * 2 = 1 + 2 - 4$$
$$= -1 \qquad \text{Hence } * \text{ is not closed}$$

    (b) For identity we need $\quad a * e = e * a = a$

$a * e = a \quad$ gives $\quad a + e - 4 = a$
$$e = 4$$

Similarly $\quad e * a = a \quad$ gives $\quad e + a - 4 = a$
$$e = 4$$

    (c) For inverse we need to find $\quad a^{-1} \quad$ such that $\quad a * a^{-1} = a^{-1} * a = e$

$a * a^{-1} = e \quad$ gives $\quad a + a^{-1} - 4 = 4$
$$a^{-1} = 8 - a$$

$a^{-1} * a = e \quad$ gives $\quad a^{-1} + a - 4 = 4$
$$a^{-1} = 8 - a$$

Only $\{0, 1, 2, 3, 4, 5, 6, 7, 8\}$ have inverses.
(Don't forget the inverse has to be a member of the set $\mathbb{N}$)

Example

Question:

The binary operation $*$ is defined by $a * b = \dfrac{a}{a+b}$ for all $a, b \in \mathbb{Z}^+$

    (a)   Determine whether or not $*$ is

        (i)   closed

        (ii) associative

        (iii) commutative

    (b)   Does $*$ have an identity element ?

    (c)   Does each element of $\mathbb{Z}^+$ have an inverse ?

Answer:

    (a) (i) $3 * 4 = \dfrac{3}{3+4}$

                $= \dfrac{3}{7}$ which is not a member of $\mathbb{Z}^+$

           Hence $*$ is not closed

(ii) If $*$ is associative then $(a * b) * c = a * (b * c)$

$$(a * b) * c = \frac{a}{a+b} * c$$

$$= \frac{\dfrac{a}{a+b}}{\dfrac{a}{a+b} + c}$$

$$= \frac{a}{a + c(a+b)}$$

$$= \frac{a}{a + ac + bc}$$

$$a * (b * c) = a * \frac{b}{b+c}$$

$$= \frac{a}{a + \dfrac{b}{b+c}}$$

$$= \frac{a(b+c)}{a(b+c) + b}$$

$$= \frac{ab + ac}{ab + ac + b}$$

Hence $*$ is not associative

Note : It is also correct to say that since $*$ is not closed then it cannot be associative.

(iii)   If * is commutative then $a * b = b * a$

$$a * b = \frac{a}{a+b}$$

$$b * a = \frac{b}{b+a}$$

Hence * is not commutative

(b)   For identity $a * e = e * a = a$

$$a * e = a \quad \text{gives} \quad \frac{a}{a+e} = a$$

$$a = a(a + e)$$

$$e = 1 - a$$

Since e has no unique value , there is no identity.

(c)   Since there is no identity, no element can have an inverse.

Exercise 4

1) The binary operation $*$ is defined as

$$a * b = 4ab - a - b \qquad \text{for all} \quad a, b \in \mathbb{R}$$

   (a) Show that $*$ is commutative

   (b) Solve $x * x = 2$

2) The binary operation $*$ is associative and commutative.
    Also $a * a = e$ and $b * b = e$

   Show that   (a) $(a * b) * (a * b) = e$

   (b) $a^3 * b^3 = ba$ (Note : $a^3 = a * a * a$)

3) The binary operation $*$ is defined as

$$a * b = a + b + 2ab \qquad \text{for all} \quad a, b \in \mathbb{R}$$

   (a) Show that $(2 * 3) * 4 = 2 * (3 * 4)$

   (b) Prove that $*$ is associative

   (c) Show that the identity element is 0

   (d) Find the inverse of $x \in \mathbb{R}$, stating any restrictions

4) The binary operation $*$ is defined as $a * b = a + b - 3$ for all $a, b \in \mathbb{N}$

   Determine whether or not $*$ is

   (a)   closed

   (b)   commutative

   (c)   associative

Groups

A set G and a binary operation $*$ are said to form a group if the following **four** conditions are met :

    1) $*$ is closed on the set G
    2) there is an identity element in set G
    3) every element in set G has an inverse
    4) $*$ is associative in set G

Example

The set $\mathbb{Z}$ and the binary operation $+$ form a group.

Proof :

1) The addition of two integers always results in an integer.
   Hence $+$ is closed on the set $\mathbb{Z}$.

2) The integer 0 is the identity element
   because $\quad a + 0 = 0 + a = a \quad$ for all integers a .

3) Every element a in the set $\mathbb{Z}$ has an inverse $-a$
   because $\quad a + -a = -a + a = 0$

4) Addition is associative

We write the group as $(\mathbb{Z}, +)$

Since $\mathbb{Z}$ is an infinite set, $(\mathbb{Z}, +)$ is said to be an infinite group.

Example

The set $G = \{1,\ i,\ -i,\ -1\}$ and the binary operation $\times$ form a group.

Proof:

Since G is a finite set we can use a table to prove that $(G,\ \times)$ is a group.

$\times$	1	i	-1	$-i$
1	1	i	-1	$-i$
i	i	-1	$-i$	1
-1	-1	$-i$	1	i
$-i$	$-i$	1	i	-1

1)   The table is closed (no new elements appear)

2)   There is an identity element (1)

3)   Every element has an inverse
     (the identity element 1 appears once in every row and once
     in every column)

4)   Multiplication is always associative.

Note: The table has symmetry about the leading diagonal.
       This indicates that the binary operation is commutative.
       In this case, the group is called an **Abelian** group.

Cayley Table

This is any table that illustrates a binary operation on a finite set.
It is not necessarily a group.

*			b	
			$\vdots$	
			$\vdots$	
a	...	...	$a*b$	

Latin Square

This is a Cayley table in which each element of the set occurs once in every row and once in every column.
Again, it is not necessarily a group.

Example:

*	1	2	3	4
1	2	1	4	3
2	4	3	2	1
3	1	4	3	2
4	3	2	1	4

Order of a group

The order of a group is the number of elements in the group.

Example :   The following group has order 3

*	I	C	D
I	I	C	D
C	C	D	I
D	D	I	C

Order of an element

The order of an element in a group is the least value of n such that $x^n = e$

(Note : $x^n = \underbrace{x * x * x * \ldots\ldots * x}_{n \text{ times}}$)

Example :   In the group above , the element C has order 3
because $C * C * C = I$

Cyclic Group

A group is said to be cyclic if the order of the group is the same as the order of
at least one element of the group.

Example

Question :

Construct a Cayley table for $G = \{2, 4, 6, 8\}$ under the binary operation of multiplication modulo 10.

(a)   Hence show that $(G, \times_{mod\ 10})$ is a group.

(b)   Write down the order of the group.

(c)   Calculate the order of each element of the group.

(d)   Hence prove that the group is cyclic.

[Note : $\times_{mod\ 10}$ is the remainder obtained when the product of two numbers is divided by 10]

Answer :

$\times_{10}$	2	4	6	8
2	4	8	2	6
4	8	6	4	2
6	2	4	6	8
8	6	2	8	4

(a)   Table is closed.
6 is the identity element.
6 occurs once in every row and once in every column , hence every element has an inverse.
Multiplication is always associative.

Hence $(G, \times_{mod\ 10})$ is a group.

(b) The group has four elements so the order of the group is 4

(c) $2 * 2 * 2 * 2 = 6$ so the order of element 2 is 4

$4 * 4 = 6$ so the order of element 4 is 2

$6 = 6$ so the order of element 6 is 1

$8 * 8 * 8 * 8 = 6$ so the order of element 8 is 4

(d) The order of element 2 (and element 8) matches the order of the group. Hence the group is cyclic.

The element 2 is called a **generator** of the group , because it can generate all of the elements of the group :

$$2^1 = 2$$
$$2^2 = 4$$
$$2^3 = 8$$
$$2^4 = 6$$

Similarly, the element 8 is also a **generator** of the group:

$$8^1 = 8$$
$$8^2 = 4$$
$$8^3 = 2$$
$$8^4 = 6$$

Note: element 8 is the inverse of element 2 (and vice-versa)

Theorem : If a is a generator of a cyclic group then a^{-1} is also a generator.

Proof : Let a be a generator of a cyclic group $(G, *)$ of order m
Then $a^m = e$
Now $a * a^{-1} = e \implies a^m * (a^{-1})^m = e$
$$\implies e * (a^{-1})^m = e$$
$$\implies (a^{-1})^m = e$$

Hence a^{-1} is a generator of $(G, *)$.

The elements of a cyclic group of order n can always be written in the form

$$\{ x, x^2, x^3, x^4, \ldots\ldots\ldots, x^n \}$$

where x is a generator of the group.
(Note: $x^n = e$)

Theorem : All cyclic groups are Abelian.

Proof : Let $(G, *)$ be a cyclic group with generator a.
Then , for any $x, y \in G$

$$x = a^m \quad \text{and} \quad y = a^n \quad \text{for some} \quad m, n \in \mathbb{Z}^+$$

Now $x * y = a^m * a^n$
$$= a^{m+n}$$
$$= a^{n+m} \quad \text{since addition is commutative}$$
$$= a^n * a^m$$
$$= y * x$$

Hence $(G, *)$ is Abelian.

Theorem :   In a group, the identity is unique

Proof :   Suppose there are two identity elements e and e' in $(G, *)$

Consider $e * e'$

If e is the identity then $e * e' = e'$

If e' is the identity then $e * e' = e$

Hence $e = e'$

Theorem :   In a group, each inverse element is unique

Proof :   Suppose the element x has two inverses x_1^{-1} and x_2^{-1} in $(G, *)$

then we have $x * x_1^{-1} = e$ and $x * x_2^{-1} = e$

Now	$x_1^{-1} = x_1^{-1} * e$	definition of identity
	$= x_1^{-1} * (x * x_2^{-1})$	from above
	$= (x_1^{-1} * x) * x_2^{-1}$	associativity
	$= e * x_2^{-1}$	definition of inverse
	$= x_2^{-1}$	definition of identity

Therefore the inverse is unique

Theorem : Given a group $(G, *)$ then for all $a, b, c \in G$

$$a * b = a * c \implies b = c \qquad \textbf{(Left Cancellation Law)}$$

Proof :

$a * b = a * c$	
$a^{-1} * (a * b) = a^{-1} * (a * c)$	every element has an inverse
$(a^{-1} * a) * b = (a^{-1} * a) * c$	associativity
$e * b = e * c$	definition of inverse
$b = c$	definition of identity

Right Cancellation Law :

Given a group $(G, *)$ then for all $a, b, c \in G$

$$b * a = c * a \implies b = c$$

The proof for the Right Cancellation Law is similar.

Sub-Groups

$(H, *)$ is a sub-group of $(G, *)$ if $H \subseteq G$ and the elements of H also form a group under the binary operation *

Consider $G = \{ e, a, b, c \}$ which has the following group table :

*	e	a	b	c
e	e	a	b	c
a	a	e	c	b
b	b	c	a	e
c	c	b	e	a

$H = \{e, a\}$ is a sub-set of G and

*	e	a
e	e	a
a	a	e

also forms a group table. Hence $(H, *)$ is a sub-group of $(G, *)$
It is also a **proper** sub-group.

$(\{e\}, *)$ is another **proper** sub-group.

$(\{e, a, b, c\}, *)$ is also a sub-group of $(G, *)$.

However, it is **not** a proper sub-group.

For example, a group of order 12 can only have sub-groups with orders 1, 2, 3, 4, 6 and 12.
A group of order 12 cannot have a sub-group of order 5.

Example

Show that the following Latin square does not represent a group :

*	e	a	b	c	d
e	e	a	b	c	d
a	a	e	c	d	b
b	b	c	d	e	a
c	c	d	a	b	e
d	d	b	e	a	c

Method 1

At first glance the table looks like it could be a group.
It is closed and it has an identity element e.
Also the identity element e appears once in every row and once in every column (indicating that every element has an inverse).
However, associativity does not work :

$$(a * b) * c = c * c$$
$$= b$$

$$a * (b * c) = a * e$$
$$= a$$

Hence the Latin square does not represent a group.

Method 2

If the Latin square represents a group then the only possible sub-groups have orders 1 and 5 (by Lagrange's Theorem).
However there is an obvious sub-group of order 2 : $(\{e, a\}, *)$
Hence the Latin square cannot be a group.

Lagrange's Theorem corollary

The order of a finite group is **divisible** by the order of any element.

Proof :   Let $(G, *)$ be a finite group of order n

If an element $x \in G$ has order 1 or n then the theorem is proved because n divides by n, and n divides by 1.

If $x \in G$ has order m where $1 < m < n$
Then $x^m = e$ and $(\{x, x^2, x^3, ..., x^m\}, *)$
is a subgroup of $(G, *)$.
The order of this subgroup is m.
By Lagranges theorem n divides by m.
Therefore the order of a finite group is divisible by the order of any element.

Permutation groups

The notation $\begin{pmatrix} 1 & 2 & 3 \\ 2 & 3 & 1 \end{pmatrix}$ indicates a mapping in which $1 \to 2$, $2 \to 3$ and $3 \to 1$

$\begin{pmatrix} 1 & 2 & 3 \\ 1 & 2 & 3 \end{pmatrix}$ indicates the identity mapping.

$\begin{pmatrix} 1 & 2 & 3 \\ 2 & 3 & 1 \end{pmatrix}\begin{pmatrix} 1 & 2 & 3 \\ 3 & 2 & 1 \end{pmatrix}$ is a binary operation which gives the result $\begin{pmatrix} 1 & 2 & 3 \\ 1 & 3 & 2 \end{pmatrix}$

The operation is performed right-to-left :
$$1 \to 3 \to 1$$
$$2 \to 2 \to 3$$
$$3 \to 1 \to 2$$

The inverse of $\begin{pmatrix} 1 & 2 & 3 \\ 2 & 3 & 1 \end{pmatrix}$ is $\begin{pmatrix} 1 & 2 & 3 \\ 3 & 1 & 2 \end{pmatrix}$ since

$$\begin{pmatrix} 1 & 2 & 3 \\ 2 & 3 & 1 \end{pmatrix}\begin{pmatrix} 1 & 2 & 3 \\ 3 & 1 & 2 \end{pmatrix} = \begin{pmatrix} 1 & 2 & 3 \\ 1 & 2 & 3 \end{pmatrix} \quad \text{and}$$

$$\begin{pmatrix} 1 & 2 & 3 \\ 3 & 1 & 2 \end{pmatrix}\begin{pmatrix} 1 & 2 & 3 \\ 2 & 3 & 1 \end{pmatrix} = \begin{pmatrix} 1 & 2 & 3 \\ 1 & 2 & 3 \end{pmatrix}$$

There are six possible permutations of the elements 1, 2, 3

Let $\quad I = \begin{pmatrix} 1 & 2 & 3 \\ 1 & 2 & 3 \end{pmatrix}$ $\qquad\qquad C = \begin{pmatrix} 1 & 2 & 3 \\ 2 & 3 & 1 \end{pmatrix}$

$\qquad A = \begin{pmatrix} 1 & 2 & 3 \\ 1 & 3 & 2 \end{pmatrix}$ $\qquad\qquad D = \begin{pmatrix} 1 & 2 & 3 \\ 3 & 1 & 2 \end{pmatrix}$

$\qquad B = \begin{pmatrix} 1 & 2 & 3 \\ 2 & 1 & 3 \end{pmatrix}$ $\qquad\qquad E = \begin{pmatrix} 1 & 2 & 3 \\ 3 & 2 & 1 \end{pmatrix}$

The six permutations can be arranged as a Cayley table:

*	I	A	B	C	D	E
I	I	A	B	C	D	E
A	A	I	D	E	B	C
B	B	C	I	A	E	D
C	C	B	E	D	I	A
D	D	E	A	I	C	B
E	E	D	C	B	A	I

The shaded square represents D * B = A

or $\quad \begin{pmatrix} 1 & 2 & 3 \\ 3 & 1 & 2 \end{pmatrix}\begin{pmatrix} 1 & 2 & 3 \\ 2 & 1 & 3 \end{pmatrix} = \begin{pmatrix} 1 & 2 & 3 \\ 1 & 3 & 2 \end{pmatrix}$

The Cayley table forms a group (a proof of associativity will not be required)

1) Table is closed.

2) Element I is the identity element.

3) The identity element I occurs once in every row and once in every column; hence every element has an inverse.

4) Associativity is given.

There are 5 proper sub-groups :

*	I	A
I	I	A
A	A	I

*	I	B
I	I	B
B	B	I

*	I	E
I	I	E
E	E	I

*	I	C	D
I	I	C	D
C	C	D	I
D	D	I	C

*	I
I	I

Note : the orders of the sub-groups are 2 , 3 and 1 which are factors of 6 (the order of the group)

Cycle notation :

$$\begin{pmatrix} 1 & 2 & 3 \\ 3 & 1 & 2 \end{pmatrix}$$ can also be written as $(1\ 3\ 2)$

This indicates a permutation cycle in which $1 \rightarrow 3$, $3 \rightarrow 2$ and $2 \rightarrow 1$

$(2\ 1\ 4\ 3)$ indicates a permutation cycle in which $2 \rightarrow 1$, $1 \rightarrow 4$, $4 \rightarrow 3$ and $3 \rightarrow 2$

$(2\ 1\ 4\ 3)$ could also be written as $(1\ 4\ 3\ 2)$ or $(4\ 3\ 2\ 1)$ because the end result

would be the same : $$\begin{pmatrix} 1 & 2 & 3 & 4 \\ 4 & 1 & 2 & 3 \end{pmatrix}$$

$(2\ 3)(4\ 1)$ indicates a permutation cycle in which $2 \rightarrow 3$, $3 \rightarrow 2$ and $4 \rightarrow 1$, $1 \rightarrow 4$

Again there are many different ways to write the cycle :

$(4\ 1)(3\ 2)$ will have the same result i.e. $$\begin{pmatrix} 1 & 2 & 3 & 4 \\ 4 & 3 & 2 & 1 \end{pmatrix}$$

$(1\ 2\ 3)(4)$ indicates a permutation cycle in which $1 \rightarrow 2$, $2 \rightarrow 3$, $3 \rightarrow 1$

and 4 is unchanged i.e. $$\begin{pmatrix} 1 & 2 & 3 & 4 \\ 2 & 3 & 1 & 4 \end{pmatrix}$$

Note: $(1\ 2\ 3\ 4) = \begin{pmatrix} 1 & 2 & 3 & 4 \\ 2 & 3 & 4 & 1 \end{pmatrix}$ however $(1)(2)(3)(4) = \begin{pmatrix} 1 & 2 & 3 & 4 \\ 1 & 2 & 3 & 4 \end{pmatrix}$

Example :

Given $P = (2\ 1\ 4\ 3)$ and $Q = (1)(3\ 4\ 2)$ calculate PQ

$$PQ = \begin{pmatrix} 1 & 2 & 3 & 4 \\ 4 & 1 & 2 & 3 \end{pmatrix} \begin{pmatrix} 1 & 2 & 3 & 4 \\ 1 & 3 & 4 & 2 \end{pmatrix}$$

$$= \begin{pmatrix} 1 & 2 & 3 & 4 \\ 4 & 2 & 3 & 1 \end{pmatrix}$$

$$= (1\ 4)(2)(3)$$

Example :

Given $P = (1\ 3\ 4\ 2)$ and $PQ = (3\ 2\ 4\ 1)$ calculate Q

$$P = \begin{pmatrix} 1 & 2 & 3 & 4 \\ 3 & 1 & 4 & 2 \end{pmatrix} \quad \Rightarrow \quad P^{-1} = \begin{pmatrix} 1 & 2 & 3 & 4 \\ 2 & 4 & 1 & 3 \end{pmatrix}$$

$$P^{-1}PQ = \begin{pmatrix} 1 & 2 & 3 & 4 \\ 2 & 4 & 1 & 3 \end{pmatrix} \begin{pmatrix} 1 & 2 & 3 & 4 \\ 3 & 4 & 2 & 1 \end{pmatrix}$$

$$I\,Q = \begin{pmatrix} 1 & 2 & 3 & 4 \\ 1 & 3 & 4 & 2 \end{pmatrix}$$

$$Q = (1)(2\ 3\ 4)$$

74

Order of elements in a permutation group

The order of element (2 3 1 5 4) is 5 since

$$\begin{pmatrix} 1 & 2 & 3 & 4 & 5 \\ 5 & 3 & 1 & 2 & 4 \end{pmatrix}^5 = \begin{pmatrix} 1 & 2 & 3 & 4 & 5 \\ 1 & 2 & 3 & 4 & 5 \end{pmatrix}$$

The order of element (1 2)(3 4 5) is 6 since

$$\begin{pmatrix} 1 & 2 & 3 & 4 & 5 \\ 2 & 1 & 4 & 5 & 3 \end{pmatrix}^6 = \begin{pmatrix} 1 & 2 & 3 & 4 & 5 \\ 1 & 2 & 3 & 4 & 5 \end{pmatrix}$$

Similarly, the order of element (2 3)(1 5 6)(7 8 4 9 10) is 30, the least common multiple of 2, 3 and 5.

Example :

Find the order of $\begin{pmatrix} 1 & 2 & 3 & 4 & 5 & 6 & 7 & 8 \\ 4 & 7 & 5 & 2 & 3 & 6 & 8 & 1 \end{pmatrix}$

$$\begin{pmatrix} 1 & 2 & 3 & 4 & 5 & 6 & 7 & 8 \\ 4 & 7 & 5 & 2 & 3 & 6 & 8 & 1 \end{pmatrix} = (1\ 4\ 2\ 7\ 8\ )(3\ 5\ )(6)$$

lcm (5, 2, 1) = 10

hence order = 10

Composition of functions as a group

Consider the three functions f, g, h such that

$$f(x) = x, \qquad g(x) = 1 - \frac{1}{x}, \qquad h(x) = \frac{1}{1-x}$$

for all real x except $\{0, 1\}$

Using composition of functions we have , for example :

$$f \circ f(x) = f(x)$$

$$g \circ h(x) = g\left(\frac{1}{1-x}\right)$$

$$= 1 - \frac{1}{\dfrac{1}{1-x}}$$

$$= 1 - (1 - x)$$

$$= x$$

$$= f(x)$$

$$h \circ h(x) = h(\frac{1}{1-x})$$

$$= \frac{1}{1 - \frac{1}{1-x}}$$

$$= \frac{1-x}{1-x-1}$$

$$= \frac{x-1}{x}$$

$$= 1 - \frac{1}{x}$$

$$= g(x)$$

Writing $f(x) = f$, $g(x) = g$, $h(x) = h$

we can form the following Cayley table :

$\circ$	f	g	h
f	f	g	h
g	g	h	f
h	h	f	g

It is easily shown that this is a group.

Geometric transformations as groups

Let I = the identity rectangle

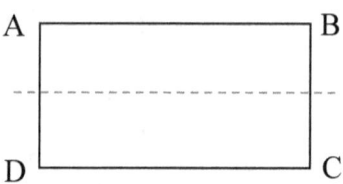

Let X = reflection in the x-axis

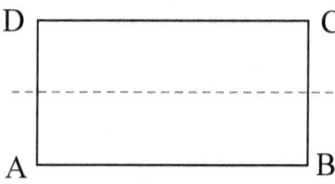

Let Y = reflection in the y-axis

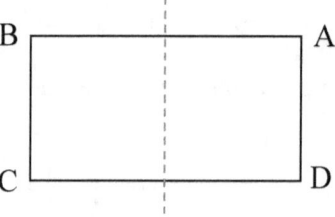

Let R = rotation 180° about O

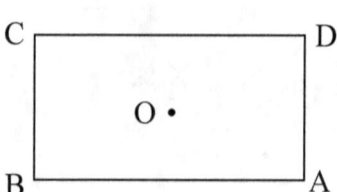

Let the binary operation $X * Y$ indicate that transformation Y is performed first and transformation X is performed second.

Hence $X * Y = R$

*	I	X	Y	R
I	I	X	Y	R
X	X	I	R	Y
Y	Y	R	I	X
R	R	Y	X	I

The resulting Cayley table is a group.

It is also an Abelian group.

Integers modulo n

$\mathbb{Z}_n$:   the set of integers modulo n

$\mathbb{Z}_n = \{ 0, 1, 2, 3, 4, \ldots\ldots\ldots, n-1 \}$

For example, $\mathbb{Z}_6 = \{ 0, 1, 2, 3, 4, 5 \}$

$(\mathbb{Z}_n, +)$ always forms a cyclic group of order n

Theorem : Given $(G, *)$ is a group and H is a non-empty subset of G then $(H, *)$ is a sub-group of $(G, *)$ if $a*b^{-1} \in H$ for all $a, b \in H$

In other words: $a*b^{-1} \in H \implies (H, *)$ is a sub-group of $(G, *)$

Proof : To be a sub-group $(H, *)$ must satisfy the 4 group requirements :

  (i) $(H, *)$ must be closed.

  (ii) $(H, *)$ must contain the identity.

  (iii) every element in $(H, *)$ must have an inverse.

  (iv) $(H, *)$ must be associative.

Identity

For all $a, b \in H$ we are given $a*b^{-1} \in H$

Since $b \in H$ we can say $b*b^{-1} \in H$

which gives $e \in H$

Inverse

For all $a, b \in H$ we are given $a*b^{-1} \in H$

Since $e \in H$ we have $e*b^{-1} \in H$

which gives $b^{-1} \in H$ for all $b \in H$

Hence each element has an inverse.

Closure

Since every element in H has an inverse in H

then b^{-1} must have an inverse in H i.e. $(b^{-1})^{-1} \in H$

Hence for all $a, b \in H$ we have $a*(b^{-1})^{-1} \in H$ or $a*b \in H$

Hence closure.

Associativity

Since associativity of $*$ applies to all elements of G, it must apply to all elements of H (as H is a subset of G)

Note: the order in which the 4 conditions are proved is important.

Theorem :   If $(G, *)$ is a finite group and H is a non-empy subset of G then $(H, *)$ is a subgroup of $(G, *)$ if $a*b \in H$ for all $a, b \in H$

In other words :   If $(H, *)$ is closed then $(H, *)$ is a subgroup

Proof :   We need to show that if $(H, *)$ is closed then

    (i)   the identity is in H

    (ii)   there is an inverse for every element in H

    (iii)   * is associative in H

Identity

As $(G, *)$ is a finite group the order of any $x \in H$ is finite.
Let this order be m where $m \in \mathbb{Z}^+$
Since $(H, *)$ is closed, this means $x^m \in H$
As $x^m = e$, this gives $e \in H$

Inverse

First, e is its own inverse.
For all other $x \in H$, $x^m = e$
Now $x^m = x^{1+m-1} = x^1 * x^{m-1}$
Similarly $x^m = x^{m-1+1} = x^{m-1} * x^1$
Hence $x^1 * x^{m-1} = x^{m-1} * x^1 = x^m$
or $x * x^{m-1} = x^{m-1} * x = e$
which gives x^{m-1} as the inverse of x.

Associativity

Since associativity of * applies to all elements of G, it must apply to all elements of H (as H is a subset of G)

Exercise 5

1) Construct a Cayley table for the set $G = \{1, 3, 7, 9\}$ under the binary operation multiplication modulo 10.

    (a) Show that the Cayley table represents a group.

    (b) Explain how the table indicates that the group is Abelian.

    (c) Use Lagrange's theorem to explain why $(G, \times_{\text{mod } 10})$ cannot have a sub-group of order 3.

    (d) Write down the proper sub-groups of $(G, \times_{\text{mod } 10})$.

    (e) Show that $(G, \times_{\text{mod } 10})$ is cyclic and state all generators.

    (f) Solve $x * 3 * x = 7$ where $x \in G$

2) (a) Show that the following are groups :

    (i) $\mathbb{Q} - \{0\}$ under multiplication.

    (ii) The set of even integers under addition.

    (iii) The set of real numbers under $*$ where $a * b = a + b - 2$ for all $a, b \in \mathbb{R}$.

(b) Explain why the following are **not** groups :

    (i) The set of odd integers under multiplication.

    (ii) $\mathbb{C}$ under multiplication.

    (iii) $\mathbb{N}$ under addition

    (iv) The set of all subsets of $\{a, b, c, d\}$ under $\cup$

3) (a) Calculate (i) $\begin{pmatrix} 1 & 2 & 3 \\ 3 & 1 & 2 \end{pmatrix}\begin{pmatrix} 1 & 2 & 3 \\ 2 & 1 & 3 \end{pmatrix}\begin{pmatrix} 1 & 2 & 3 \\ 3 & 2 & 1 \end{pmatrix}$

(ii) $\begin{pmatrix} 1 & 2 & 3 & 4 \\ 3 & 1 & 4 & 2 \end{pmatrix}\begin{pmatrix} 1 & 2 & 3 & 4 \\ 2 & 3 & 4 & 1 \end{pmatrix}$

(b) Find the inverse of $\begin{pmatrix} 1 & 2 & 3 & 4 \\ 3 & 1 & 4 & 2 \end{pmatrix}$

(c) Find the order of $\begin{pmatrix} 1 & 2 & 3 & 4 \\ 3 & 1 & 4 & 2 \end{pmatrix}$

(d) Given $A = \begin{pmatrix} 1 & 2 & 3 & 4 \\ 3 & 1 & 4 & 2 \end{pmatrix}$ and $B = \begin{pmatrix} 1 & 2 & 3 & 4 \\ 4 & 2 & 1 & 3 \end{pmatrix}$

solve $AX = B$

4) $p = \begin{pmatrix} 1 & 2 & 3 & 4 & 5 \\ 2 & 3 & 1 & 5 & 4 \end{pmatrix}$ is a generator of a cyclic group $(G , \circ)$

where $\circ$ represents composition of permutations.

(a) Write down the identity element in $(G , \circ)$

(b) Find (i) the order of $(G , \circ)$

(ii) $p \circ p$

(iii) the inverse of p

5) Given $q = (1\ 3\ 7)(2\ 4\ 6\ 8\ 10)(5\ 9)$

Find   (a) the order of q

(b) q^3

6) Set A consists of all possible permutations of the integers $1, 2, 3, 4$.
Given $p, q \in A$ where $p = (1\ 2\ 4)$ and $q = (1\ 3)(2\ 4)$

(a) Find   (i) p^{-1}
(ii) $p \circ q$

(b) State the order of   (i) p
(ii) q

Note : $p = (1\ 2\ 4)$ can also be written as $p = (1\ 2\ 4)(3)$

7) Let P be the cycle $(1\ 3\ 2\ 4)$ and Q be the cycle $(1\ 2)(3\ 4)$

(i) Calculate PQ

(ii) Let $I = (1)(2)(3)(4)$ and $R = PQ$

Show that $\{\ I, P, Q, R\ \}$ forms a group under composition of mappings.
You are given that composition of mappings is associative.

84

8) (a) Construct a Cayley table for $\mathbb{Z}_6$ under addition modulo 6

(b) Show that the Cayley table represents a group.

(c) Prove that the group is cyclic and write down all of the generators.

(d) Write down all proper sub-groups.

9) $(G , *)$ is an Abelian group of order 6.
Four of the elements in the group are e, x, y and xy^2
where e is the identity element and $xy^2 = x * y * y$.
Given that element x has order 2 and element y has order 3

(a) Write down the missing two elements of the group

(b) Show that $x^3 y^4 = yx$

(c) Show that the order of xy^2 is 6.

(d) Explain why $(G , *)$ is cyclic.

10) (a) Show that $\mathbb{Z}_5 - \{0\}$ forms a cyclic group under multiplication modulo 5.

(b) (i) Construct a Cayley table for set $\mathbb{Z}_6 - \{0\}$ under multiplication modulo 6.

(ii) Hence show that $\mathbb{Z}_6 - \{0\}$ does **not** form a group under multiplication modulo 6.

11) (a) Prove that $[a*b^{-1}]^{-1} = b*a^{-1}$

(b) Let $(H, *)$ be a subgroup of $(G, *)$ and let R be a relation
defined on G by $aRb \Leftrightarrow a*b^{-1} \in H$.
Prove that R is an equivalence relation.

12) Prove that if a cyclic group has only one generator then it **cannot** have
more than two elements.

13) Set G has elements of the form $\dfrac{1+2x}{1+2y}$ where $x \in \mathbb{Z}$ and $y \in \mathbb{Z}$.

Show that G forms a group under multiplication.

Cosets

Let $(H, *)$ be a subgroup of $(G, *)$ and let $x \in G$.

Then a **left** coset of $(H, *)$ is defined by $xH = \{x*h : h \in H\}$

and a **right** coset of $(H, *)$ is defined by $Hx = \{h*x : h \in H\}$

Example: Consider the group $(G, *)$ shown below

*	e	a	b	c	d	f
e	e	a	b	c	d	f
a	a	b	e	f	c	d
b	b	e	a	d	f	c
c	c	d	f	e	a	b
d	d	f	c	b	e	a
f	f	c	d	a	b	e

Let $(H, *)$ be the subgroup $(\{e, a, b\}, *)$

Then the six possible **left** cosets of $(H, *)$ are :

$$eH = \{e*e, \ e*a, \ e*b\} = \{e, a, b\}$$

$$aH = \{a*e, \ a*a, \ a*b\} = \{a, b, e\}$$

$$bH = \{b*e, \ b*a, \ b*b\} = \{b, e, a\}$$

$$cH = \{c*e, \ c*a, \ c*b\} = \{c, d, f\}$$

$$dH = \{d*e, \ d*a, \ d*b\} = \{d, f, c\}$$

$$fH = \{f*e, \ f*a, \ f*b\} = \{f, c, d\}$$

Note : the first 3 cosets are the same, as are the second 3.
Hence the left cosets are $\{e, a, b\}$ and $\{c, d, f\}$

The six possible **right** cosets of $(H, *)$ are :

$$He = \{e*e,\ a*e,\ b*e\} \ = \ \{e, a, b\}$$

$$Ha = \{e*a,\ a*a,\ b*a\} \ = \ \{a, b, e\}$$

$$Hb = \{e*b,\ a*b,\ b*b\} \ = \ \{b, e, a\}$$

$$Hc = \{e*c,\ a*c,\ b*c\} \ = \ \{c, f, d\}$$

$$Hd = \{e*d,\ a*d,\ b*d\} \ = \ \{d, c, f\}$$

$$Hf = \{e*f,\ a*f,\ b*f\} \ = \ \{f, d, c\}$$

These simplify to $\{e, a, b\}$ and $\{c, d, f\}$

In this case the **left** cosets and the **right** cosets are the same. This is not always the case.

Homomorphism

Let $(G, *)$ and $(H, \bullet)$ be two groups.

The function $f : G \to H$ is a homomorphism if

$$f(a * b) = f(a) \bullet f(b) \qquad \text{for all } a, b \in G$$

Example 1 :

Given two groups $(\mathbb{R}, +)$ and $(\mathbb{R} - \{0\}, \times)$

The function $f : \mathbb{R} \to \mathbb{R} - \{0\}$ where $f(x) = e^x$ is a homomorphism

because $f(x + y) = f(x) \times f(y)$ for all $x, y \in \mathbb{R}$

This can easily be shown by simple algebra : $f(x + y) = e^{x+y}$

$$= e^x \times e^y$$

$$= f(x) \times f(y)$$

Example 2 :

Given two groups $(\mathbb{Z}, +)$ and $(\mathbb{Z}_3, +)$

The function $f : \mathbb{Z} \to \mathbb{Z}_3$ where $f(x) = x \bmod 3$ is a homomorphism

because $f(x + y) = f(x) + f(y)$ for all $x, y \in \mathbb{Z}$

e.g. $f(4 + 7) = f(4) + f(7)$

$$f(11) = f(4) + f(7)$$

$$2 = 1 + 1$$

Example 3 :

Given the identical groups $(\mathbb{Z}, +)$ and $(\mathbb{Z}, +)$
prove that $f: \mathbb{Z} \to \mathbb{Z}$, $f(x) = x^2$ is **not** a homomorphism

For a homomorphism we need $f(x + y) = f(x) + f(y)$ for all $x, y \in \mathbb{Z}$

A simple counter-example is enough proof :

Consider $f(2 + 3) = f(2) + f(3)$

$$\text{LHS} = f(2 + 3)$$

$$= f(5)$$

$$= 25$$

$$\text{RHS} = f(2) + f(3)$$

$$= 4 + 9$$

$$= 13$$

Since the results are not equal we do not have a homomorphism

Note : LHS = left hand side

RHS = right hand side

Kernal

We define the kernal of f to be the set of elements in G
which are mapped onto the identity element in H.

In **Example 1** the kernal is { 0 } because $f(0) = e^0 = 1$

In other words, 0 in set $\mathbb{R}$ maps onto 1 (the identity in $\mathbb{R} - \{0\}$)

In **Example 2** the kernal is { .... -6 , -3 , 0 , 3 , 6 , 9 , 12 , ...}

because $\qquad f(-6) = 0$

$$f(-3) = 0$$

$$f(0) = 0$$

$$f(3) = 0$$

$$f(6) = 0$$

$$f(9) = 0$$

$$f(12) = 0$$

etc.

These are all of the elements of $\mathbb{Z}$ which map onto 0
(the identity in $\mathbb{Z}_3$)

The kernel of f is written as Ker (f)

Ker (f) is always a subset of G.

Further examples :

Question:

Given two groups $(\mathbb{Z}, +)$ and $(e, \times)$

    (i) Show that $f : \mathbb{Z} \rightarrow e$, $f(x) = e$ is a homomorphism.

    (ii) State the kernal of f

Answer:

    (i) $f(x + y) = e$
$$= e \times e$$
$$= f(x) \times f(y)$$

        Hence f is a homomorphism

    (ii) The kernal of f is the entire set $\mathbb{Z}$ since all elements in $\mathbb{Z}$ are mapped onto the identity.

        This is called the trivial homomorphism (for obvious reasons)

Question:

Given two groups $(\mathbb{R} - \{0\}, \times)$ and $(\mathbb{R} - \{0\}, \times)$

    (i) Show that $f : \mathbb{R} - \{0\} \rightarrow \mathbb{R} - \{0\}$, $f(x) = |x|$ is a homomorphism.

    (ii) State the kernal of f

Answer:

    (i) $f(x \times y) = |x \times y|$
$$= |x| \times |y|$$
$$= f(x) \times f(y)$$

        Hence f is a homomorphism.

(ii) The kernal of f is $\{-1, 1\}$

since $f(-1) = |-1|$

$= 1$

and $f(1) = |1|$

$= 1$

Question:

Let $(G, *)$ be a group.

Define a mapping $f: G \rightarrow G$ where $f(x) = x^2$ for all $x \in G$

Show that f is a homomorphism if and only if G is Abelian.

Answer:

Homomorphism	$\Leftrightarrow$	$f(a * b) = f(a) * f(b)$
	$\Leftrightarrow$	$(a * b)^2 = a^2 * b^2$
	$\Leftrightarrow$	$a\,b\,a\,b = a\,a\,b\,b$
	$\Leftrightarrow$	$a^{-1} a b a b b^{-1} = a^{-1} a a b b b^{-1}$
	$\Leftrightarrow$	$e\,b\,a\,e = e\,a\,b\,e$
	$\Leftrightarrow$	$b\,a = a\,b$
	$\Leftrightarrow$	G is Abelian

Isomorphism

If a homomorphism is bijective then it is called an isomorphism.

Let $(G, *)$ and $(H, \bullet)$ be two groups.

Then the function $f : G \to H$ is an isomorphism if

$$f(a * b) = f(a) \bullet f(b) \quad \text{for all} \quad a, b \in G$$

and

$$f(a) = f(b) \Rightarrow a = b$$

Example

Question:

Given two groups $(\mathbb{R}, +)$ and $(\mathbb{R} - \{0\}, \times)$

Show that $f : \mathbb{R} \to \mathbb{R} - \{0\}$, $f(x) = e^x$ is an isomorphism.

Answer:

$$\begin{aligned} f(x + y) &= e^{x+y} \\ &= e^x \times e^y \\ &= f(x) \times f(y) \end{aligned}$$

Hence f is a homomorphism.

$$f(x) = f(y) \Rightarrow e^x = e^y$$

$$\Rightarrow x = y$$

Hence f is bijective.

Therefore f is an isomorphism.

An isomorphism is a one-to-one correspondence between two groups.

To prove that an isomorphism exists between two finite groups
it is sufficient to show this one-to-one correspondence.

Example :

Consider the two groups shown in the following Cayley tables

$\times$	1	i	-1	$-i$
1	1	i	-1	$-i$
i	i	-1	$-i$	1
-1	-1	$-i$	1	i
$-i$	$-i$	1	i	-1

$*$	e	a	b	c
e	e	a	b	c
a	a	b	c	e
b	b	c	e	a
c	c	e	a	b

It can easily be seen that $1 \leftrightarrow e$

$i \leftrightarrow a$

$-1 \leftrightarrow b$

$-i \leftrightarrow c$

Hence the two groups are isomorphic.

Isomorphic groups have the same properties.
If one group is cyclic then so is the other.
Also, corresponding elements will have the same order.

In the example shown $i^4 = 1$ and $a^4 = e$
(the two corresponding elements are both generators of their respective groups)

Also $(-1)^2 = 1$ and $b^2 = e$
(the two corresponding elements both have order 2)

Using $f(x \times y) = f(x) * f(y)$ for all $x, y \in G$

We have, for example, $f(-i \times i) = f(-i) * f(i)$

$$f(1) = f(-i) * f(i)$$

$$= c * a$$

$$= e$$

The bijection is $f(1) = e$

$$f(i) = a$$

$$f(-1) = b$$

$$f(-i) = c$$

Consider the two groups represented by the following two Caley tables :

$+_4$	0	1	2	3
0	0	1	2	3
1	1	2	3	0
2	2	3	0	1
3	3	0	1	2

$\times_{10}$	2	4	6	8
2	4	8	2	6
4	8	6	4	2
6	2	4	6	8
8	6	2	8	4

At first glance , the two tables do not follow the same pattern. However , simple re-arrangement of the second table results in two possible isomorphisms :

$\times_{10}$	6	2	4	8
6	6	2	4	8
2	2	4	8	6
4	4	8	6	2
8	8	6	2	4

$\times_{10}$	6	8	4	2
6	6	8	4	2
8	8	4	2	6
4	4	2	6	8
2	2	6	8	4

$0 \leftrightarrow 6$ $0 \leftrightarrow 6$

$1 \leftrightarrow 2$ $1 \leftrightarrow 8$

$2 \leftrightarrow 4$ $2 \leftrightarrow 4$

$3 \leftrightarrow 8$ $3 \leftrightarrow 2$

Useful property :

$(\mathbb{Z}_n , +)$ is always isomorphic to any other cyclic group of order n

---

Theorem : If two groups $(G, *)$ and $(H, \bullet)$ are homomorphic then the identity of $(G, *)$ is mapped onto the identity of $(H, \bullet)$

Proof : Let e_G be the identity element of $(G, *)$

and let $f : G \to H$ be the homomorphism

For all $a, b \in G$ we have $f(a*b) = f(a) \bullet f(b)$

Hence $f(a*e_G) = f(a) \bullet f(e_G)$

and $f(a) = f(a) \bullet f(e_G)$

Similarly $f(e_G * a) = f(e_G) \bullet f(a)$

$f(a) = f(e_G) \bullet f(a)$

Combining gives

$f(a) \bullet f(e_G) = f(e_G) \bullet f(a) = f(a)$

Therefore $f(e_G)$ is e_H , the identity element of $(H, \bullet)$

Theorem : If two groups $(G, *)$ and $(H, \bullet)$ are homomorphic then the inverse of an element in $(G, *)$ is mapped onto the inverse of the corresponding element in $(H, \bullet)$

Proof : Let e_G be the identity element of $(G, *)$ and

let $f : G \to H$ be the homomorphism

For all $a, b \in G$ we have $f(a*b) = f(a) \bullet f(b)$

Hence $f(a*a^{-1}) = f(a) \bullet f(a^{-1})$

and $f(e_G) = f(a) \bullet f(a^{-1})$

Similarly $f(a^{-1}*a) = f(a^{-1}) \bullet f(a)$

$f(e_G) = f(a^{-1}) \bullet f(a)$

Combining gives

$$f(a) \bullet f(a^{-1}) = f(a^{-1}) \bullet f(a) = f(e_G)$$

Since $f(e_G)$ is e_H , the identity element of $(H, \bullet)$ we have $f(a^{-1})$ as the inverse of $f(a)$

Hence $f(a^{-1}) = [f(a)]^{-1}$

Theorem :   Let $f : G \rightarrow H$ be a homomorphism from group $(G, *)$ to group $(H, \bullet)$.

Then the kernal of f is a subgroup of $(G, *)$

Proof :   We need to show that the set $\operatorname{Ker} f$ has the following properties :

1) the set is closed under *

2) the set contains the identity element e_G

3) every element in the set has an inverse

4) the set is associative under *

1) *Closure*

Suppose $x, y \in \operatorname{Ker} f$

Then $f(x) = f(y) = e_H$
where e_H is the identity element of group $(H, \bullet)$

Now $f(x * y) = f(x) \bullet f(y)$ for all $x, y \in \operatorname{Ker} f$

$$= e_H \bullet e_H$$

$$= e_H$$

Hence $x * y \in \operatorname{Ker} f$

Therefore closure.

2) *Identity*

Let e_G be the identity element of $(G, *)$

Since $f(a * b) = f(a) \bullet f(b)$ for all $a, b \in (G, *)$

we have $f(a * e_G) = f(a) \bullet f(e_G)$

$$f(a) = f(a) \bullet f(e_G)$$

Let $x \in \operatorname{Ker} f$

Then $f(x) = f(x) \bullet f(e_G)$ since x is also an element of G

$$e_H = e_H \bullet f(e_G)$$

$$e_H = f(e_G)$$

$$f(e_G) = e_H$$

Hence $e_G \in \operatorname{Ker} f$

3) *Inverse*

Let a^{-1} be the inverse of element a in $(G, *)$

Since $f(a * b) = f(a) \bullet f(b)$ for all $a, b \in (G, *)$

we have $f(a * a^{-1}) = f(a) \bullet f(a^{-1})$

$$f(e_G) = f(a) \bullet f(a^{-1})$$

$$e_H = f(a) \bullet f(a^{-1})$$

Let $x \in \operatorname{Ker} f$

Then $e_H = f(x) \bullet f(x^{-1})$

$$= e_H \bullet f(x^{-1})$$

$$= f(x^{-1})$$

$$f(x^{-1}) = e_H \quad \Rightarrow \quad x^{-1} \in \operatorname{Ker} f$$

4) *Associativity*

Since $\operatorname{Ker} f$ is a subset of G, the elements of $\operatorname{Ker} f$
are associative under $*$

Hence $(\operatorname{Ker} f, *)$ is a sub-group of $(G, *)$

Theorem : Let $f : G \rightarrow H$ be a homomorphism from group $(G, *)$
to group $(H, \bullet)$.

Then the range of f is a subgroup of $(H, \bullet)$

Proof : Let the range of the homomorphism be $\text{Im}(f)$

We need to show that the set $\text{Im}(f)$ has the following
properties :

    1) the set is closed under $\bullet$

    2) the set contains the identity element e_H

    3) every element in the set has an inverse

    4) the set is associative under $\bullet$

1) *Closure*

    Let $x, y \in (G, *)$

    Then $f(x)$ and $f(y)$ both belong to $\text{Im}(f)$

    Likewise $f(x * y)$ belongs to $\text{Im}(f)$

    But $f(x * y) = f(x) \bullet f(y)$ for all $x, y \in (G, *)$

    Therefore $f(x) \bullet f(y)$ also belongs to $\text{Im}(f)$

    Hence closure.

2) *Identity*

Let e_G be the identity element of $(G, *)$

Since $f(e_G) = e_H$ we have $e_H \in \text{Im}(f)$

3) *Inverse*

$x \in (G, *) \quad \Rightarrow \quad x^{-1} \in (G, *)$

Hence $f(x^{-1}) \in \text{Im}(f)$

Since $f(x^{-1}) = [f(x)]^{-1}$ we have $[f(x)]^{-1} \in \text{Im}(f)$

So the inverse of every element in the range is also in the range.

4) *Associativity*

Since $\text{Im}(f)$ is a subset of H, the elements of $\text{Im}(f)$ are associative under $\bullet$

Hence $(\text{Im}(f), \bullet)$ is a sub-group of $(H, \bullet)$

Exercise 6

1)   (a) Solve $z^5 = 1$ giving your answers in the form $cis\theta$ where $0 \leq \theta \leq 2\pi$

     (b) Show that the five answers to part (a) form a group under multiplication.

     (c) Show that the group in part (b) is isomorphic to $(\mathbb{Z}_5, +)$

2)   Given the groups $(\mathbb{R} - \{0\}, \times)$ and $(\mathbb{R}, +)$

     (i) Show that $f : \mathbb{R} - \{0\} \to \mathbb{R}$, $f(x) = \log|x|$ is a homomorphism

     (ii) Write down the kernal of f

3)   (a) Construct Cayley tables for $(\mathbb{Z}_4, +)$ and $(\mathbb{Z}_5 - \{0\}, \times)$

     (b) Hence show that the two groups are isomorphic.

4)   Show that the mapping $f : x \to 2^x$ from the set of integers $\mathbb{Z}$ to the set $G = \{......\dfrac{1}{4}, \dfrac{1}{2}, 1, 2, 4, 8, ...\}$ sets up an isomorphism between groups $(\mathbb{Z}, +)$ and $(G, \times)$

5)   Given the groups $(\mathbb{Z}, +)$ and $(\{1, i, -1, -i\}, \times)$

     (i) Show that $f : \mathbb{Z} \to \{1, i, -1, -i\}$, $f(n) = i^n$ is a homomorphism

     (ii) Write down the kernal of f

6) Given the groups $(\mathbb{Z}, +)$ and $(\mathbb{Z}, +)$
   Show that that $f : \mathbb{Z} \to \mathbb{Z}$, $f(n) = n + 1$ is **not** a homomorphism.

7) Let $(G, *)$ be a group.
   Define a mapping $f : G \to G$ where $f(x) = x^{-1}$ for all $x \in G$
   Show that f is a homomorphism of groups if and only if $(G, *)$
   is Abelian.

8) Given the group $(G, +_6)$ shown below

$+_6$	0	1	2	3	4	5
0	0	1	2	3	4	5
1	1	2	3	4	5	0
2	2	3	4	5	0	1
3	3	4	5	0	1	2
4	4	5	0	1	2	3
5	5	0	1	2	3	4

(a) Find a subgroup of order 2.

(b) Prove that $(\{0, 2, 4\}, +_6)$ is a subgroup.

(c) Calculate the left cosets of $(\{0, 2, 4\}, +_6)$

Answers

Exercise 1

1) (i) { 2 , 3 , 5 }

   (ii) { 4 , 6 , 10 }

   (iii) { 1 , 2 , 3 , 4 , 5 , 6 , 8 , 9 , 10 }

   (iv) { 7 }

   (v) { 1 , 7 , 8 , 9 }

2) (i) { 2 , 3 , 4 , 5 , 6 , 7 , 8 , 9 , 10 }

   (ii) { 1 , 2 , 3 , 4 }

   (iii) { 2 , 3 , 4 }

   (iv) { 5 , 6 , 7 , 8 , 9 , 10 }

   (v) { 11 , 12 , 13 , 14 , 15 , 16 , 17 , 18 , 19 }

   (vi) { 1 , 5 , 6 , 7 , 8 , 9 , 10 }

3) 7

4) (i) $\mathbb{Z}$

   (ii) $\mathbb{Z}^-$

   (iii) U

   (iv) 0

   (v) 1

5) $A \cap (A \cap B')'$ $=$ $A \cap (A' \cup B)$ *DeMorgan's rule*

$=$ $(A \cap A') \cup (A \cap B)$ *Distributive rule*

$=$ $\varnothing \cup (A \cap B)$ *Intersection rule*

$=$ $A \cap B$ *Union rule*

6) $(A \cap B)' \cap (A \cup B) = (A' \cup B') \cap (A \cup B)$ *DeMorgan's rule*

$=$ $(A' \cap A) \cup (A' \cap B) \cup (B' \cap A) \cup (B' \cap B)$ *Distributive rule*

$=$ $\varnothing \cup (A' \cap B) \cup (B' \cap A) \cup \varnothing$ *Intersection rule*

$=$ $(A' \cap B) \cup (B' \cap A)$ *Union rule*

$=$ $(B - A) \cup (A - B)$

$=$ $A \triangle B$

8)   Let $x \in A$
Then $x \in A \cap B$ since $A \cap B = A$
Hence $x \in B$
Which gives $A \subseteq B$

9)   (i) $(1, 4)$ $(1, 5)$ $(2, 4)$ $(2, 5)$ $(3, 4)$ $(3, 5)$

(ii) $(4, 1)$ $(4, 2)$ $(4, 3)$ $(5, 1)$ $(5, 2)$ $(5, 3)$

10) $(A\cap B) - (A\cap C) = (A\cap B) \cap (A\cap C)'$

$\qquad\qquad\qquad\qquad = (A\cap B) \cap (A'\cup C')$

$\qquad\qquad\qquad\qquad = [(A\cap B) \cap A'] \cup [(A\cap B) \cap C']$

$\qquad\qquad\qquad\qquad = [(A\cap A' \cap B] \cup [(A\cap B) \cap C']$

$\qquad\qquad\qquad\qquad = [\varnothing \cap B] \cup [A\cap(B \cap C')]$

$\qquad\qquad\qquad\qquad = \varnothing \cup [A\cap(B-C)\,]$

$\qquad\qquad\qquad\qquad = A\cap(B-C)$

Exercise 2

1)   (a) $\{3, 8, 9\}$

      (b) $y > 0$

      (c) $y > 1$

      (d) $y \ge -1$

      (e) $\mathbb{R} - \{3\}$

2)   (a) $f^{-1}(x) = \dfrac{3x+1}{x-2}$, $x \in \mathbb{R} - \{2\}$

      (b) $f^{-1}(x) = \dfrac{x^2 - 8}{2}$, $x \ge 0$

      (c) $f^{-1}(x) = \dfrac{3 \pm \sqrt{1+4x}}{2}$, $x \ge -\dfrac{1}{4}$

      (d) $f^{-1}(x) = \ln\left(\dfrac{-1+\sqrt{4x-3}}{2}\right)$, $x > 1$

3)   (a)   (i) $f(a) = f(b) \Rightarrow e^a = e^b$
$$\Rightarrow a = b$$

(ii)   range of $e^x \neq$ codomain $\mathbb{R}$
(or negative numbers and zero in the codomain have no corresponding elements in the domain. For example, there is no value of x such that $e^x = -8$)

(iii) $\log_e x$

4)   (i) $f(a) = f(b) \Rightarrow a^3 = b^3$
$$\Rightarrow \quad a = b \quad \text{hence injection}$$

(ii)   range of x^3 = codomain $\mathbb{R}$
(or all elements n in the codomain have corresponding elements $\sqrt[3]{n}$ in the domain)

Hence surjective.

Injection + Surjection $\Rightarrow$ bijection

5) $y \geq -1$

6)   (a) $(8, 2)$

(b) $(3, 2)$

(c) $f^{-1}(x, y) = \left(\dfrac{x + 3y}{14}, \dfrac{2x - y}{7} \right)$

(d)　$f(a,b) = f(c,d) \Rightarrow (2a+3b \, , \, 4a-b) = (2c+3d \, , \, 4c-d)$

$$\Rightarrow \quad 2a+3b = 2c+3d$$
$$4a-b = 4c-d$$

$$\Rightarrow \quad 2a+3b = 2c+3d$$
$$12a-3b = 12c-3d$$

Adding gives $14a = 14c$ or $a = c$

Substituting gives $b = d$

Hence $(a,b) = (c,d) \Rightarrow$ injection

Every element (m, n) in the co-domain corresponds to
element $(\dfrac{m+3n}{14}, \dfrac{2m-n}{7})$ in the domain.
Hence $f(x,y)$ is surjective.

Injection + Surjection $\Rightarrow$ bijection

Exercise 3

1) (a)　Any triangle is similar to itself ; hence R is reflexive.

If triangle a is similar to triangle b then triangle b is similar to triangle a ; hence R is symmetric.

If triangle a is similar to triangle b and triangle b is similar to triangle c then triangle a is similar to triangle c ; hence R is transitive.

Reflexive + Transitive + Symmetric $\Rightarrow$ Equivalence relation

(b)　Sets of similar triangles

2) (a) $\{ (1, 1)\ (2, 2)\ (5, 5)\ (2, 1)\ (3, 2)\ (5, 4)\ (1, 3)\ (3, 1) \}$

(b) $\{ (1, 2)\ (2, 3)\ (3, 3) \}$ and $\{ (4, 4)\ (4, 5) \}$

3) (a) $aRb \iff a^2 \equiv b^2 \pmod 2$

$\iff a^2 - b^2 = 2m$ \qquad where $m \in \mathbb{Z}$

aRa is valid because $a^2 - a^2 = 2m$ is true (for $m = 0$)

Hence R is reflexive.

$aRb \iff a^2 - b^2 = 2m$
$\Rightarrow b^2 - a^2 = -2m$
$\Rightarrow b^2 - a^2 = 2n$ \qquad where $n \in \mathbb{Z}$ (and $n = -m$)
$\Rightarrow bRa$

Hence R is symmetric.

$aRb \iff a^2 - b^2 = 2m$ \quad and \quad $bRc \iff b^2 - c^2 = 2n$
$\Rightarrow a^2 - b^2 + b^2 - c^2 = 2m + 2n$
$\Rightarrow a^2 - c^2 = 2(m + n)$
$\Rightarrow a^2 - c^2 = 2p$ \quad where $p \in \mathbb{Z}$ (and $p = m + n$)
$\Rightarrow aRc$

Hence R is transitive.

Reflexive + Transitive + Symmetric $\Rightarrow$ Equivalence relation

(b) $\{0, 2, 4, 6, 8, \ldots\ldots\}$ and $\{1, 3, 5, 7, \ldots\ldots\ldots\}$

4)   Not transitive: $ab \geq 0$ and $bc \geq 0$ does not necessarily mean $ac \geq 0$

For example $a = -4$, $b = 0$, $c = 2$

5)   (a) $aRa \Leftrightarrow a + a = 2n$
$\Rightarrow 2a = 2n$

Hence R is reflexive.

$aRb \Leftrightarrow a + b = 2n$
$\Rightarrow b + a = 2n$
$\Rightarrow bRa$

Hence R is symmetric.

$aRb \Leftrightarrow a + b = 2n$ and $bRc \Leftrightarrow b + c = 2m$
$\Rightarrow a + b + b + c = 2n + 2m$
$\Rightarrow a + c = 2n + 2m - 2b$
$\Rightarrow a + c = 2p$
$\Rightarrow aRc$

Hence R is transitive.

Reflexive + Transitive + Symmetric $\Rightarrow$ Equivalence relation

(b) $\{ .... -4, -2, 0, 2, 4, ... \}$ and $\{ .... -3, -1, 1, 3, ... \}$

Exercise 4

1) (a) $b * a$ $= 4ba - b - a$
                    $= 4ab - a - b$
                    $= a * b$

       Hence $*$ is commutative

   (b) $x = -\dfrac{1}{2}$, 1

2) (a) $(a * b) * (a * b) = a * (b * a) * b$ associativity

                          $= a * (a * b) * b$ commutivity

                          $= (a * a) * (b * b)$ associativity

                          $= e * e$ given

                          $= e$ definition of identity

   (b) $a^3 * b^3 = a^2 * a * b * b^2$

                   $= e * a * b * e$ given

                   $= a * b$ definition of identity

                   $= b * a$ commutativity

3) (a) $(2 * 3) * 4 = (2+3+12) *4$

$$= 17*4$$

$$= 17+4+136$$

$$= 157$$

$2 * (3 * 4) = 2*(3+4+24)$

$$= 2*31$$

$$= 2+31+124$$

$$= 157$$

(b) $(a * b) * c = (a + b + 2ab) * c$

$$= (a + b + 2ab) + c + 2(a + b + 2ab) c$$

$$= a + b + c + 2ab + 2ac + 2bc + 4abc$$

$a * (b * c) = a * (b + c + 2bc)$

$$= a + (b + c + 2bc) + 2a (b + c + 2bc)$$

$$= a + b + c + 2bc + 2ab + 2ac + 4abc$$

$(a * b) * c = a * (b * c) \Rightarrow$ associativity

(c)   Method 1 : $a * 0 = a + 0 + 2(a)(0)$

$$= a$$

$$0 * a = 0 + a + 2(0)(a)$$

$$= a$$

Hence the identity element is 0

Method 2 : $a * e = a$

$$a + e + 2ae = a$$

$$e + 2ae = 0$$

$$e(1 + 2a) = 0$$

$$e = 0$$

Similarly , $e * a = a$ will give the same result
(since $*$ is commutative)

(d) $a * a^{-1} = e$

$$a + a^{-1} + 2a\,a^{-1} = 0$$

$$a^{-1}(1 + 2a) = -a$$

$$a^{-1} = \frac{-a}{1 + 2a}$$

$a^{-1} * a = e$ will give the same result (since $*$ is commutative)

Restriction: $a \neq -\dfrac{1}{2}$

4) (a) Since $1 * 1 = 1 + 1 - 3$

$$= -1$$

and $-1 \notin \mathbb{N}$ we have shown that $*$ is not closed

(b) $b * a = b + a - 3$

$$= a + b - 3$$

$$= a * b$$

Hence $*$ is commutative .

(c) As $*$ is not closed, associativity is not possible.

Exercise 5

1)

$\times_{10}$	1	3	7	9
1	1	3	7	9
3	3	9	1	7
7	7	1	9	3
9	9	7	3	1

(a) Table is closed.
Identity element is 1.
Identity element, 1 , appears once in every row
and once in every column; therefore every element
has a unique inverse.
Multiplication is associative.
Hence the table forms a group.

(b)   Symmetry about the leading diagonal

(c)   Order of the group is 4 $\Rightarrow$ only possible sub-groups
have orders 1, 2 or 4.
(alternatively: 3 is not a factor of 4)

(d)

$\times_{10}$	1
1	1

$\times_{10}$	1	9
1	1	9
9	9	1

(e) $3^1 = 3$
$3^2 = 9$
$3^3 = 7$
$3^4 = 1$

Order of element 3 is 4 , and order of group = 4
Hence cyclic.
Generators are 3 and 7

(f) $x = 7$

2) (a) (i) $\dfrac{a}{b} \times \dfrac{c}{d} = \dfrac{ac}{bd}$ as long as $b \neq 0$ and $d \neq 0$

(rational $\times$ rational = rational). Hence closed.

Identity element = 1
Every element x has an inverse $\dfrac{1}{x}$ since $x \neq 0$

Multiplication is associative.

(ii)　　$2a + 2b = 2(a + b)$　　hence closure.

Identity element $= 0$

$2x + (-2x) = 0$　　hence every element has an inverse.

Addition is associative.

(iii)　Closed :　addition and subtraction of real numbers will always result in a real number.

Identity element :　　　　　$a * e = a$

$$a + e - 2 = a$$

$$e = 2$$

since * is commutative we also have $e * a = a$

Inverse :　　　　　$a * a^{-1} = 2$

$$a + a^{-1} - 2 = 2$$

$$a^{-1} = 4 - a$$

since * is commutative , $a^{-1} * a = 2$ will give the same result.

Associativity :　　　　$(a * b) * c = (a + b - 2) * c$

$$= (a + b - 2) + c - 2$$

$$= a + b + c - 4$$

$$a * (b * c) = a * (b + c - 2)$$

$$= a + (b + c - 2) - 2$$

$$= a + b + c - 4$$

since $(a * b) * c = a * (b * c)$ we have associativity

(b) (i) No inverse : $a^{-1} = \dfrac{1}{a}$ and $\dfrac{1}{a} \notin \mathbb{Z}$

    (ii) Zero has no inverse

    (iii) No inverse : $a^{-1} = -a$ and $-a \notin \mathbb{N}$

    (iv) Identity $= \varnothing$ so no inverse possible.

3) (a) (i) $\begin{pmatrix} 1 & 2 & 3 \\ 2 & 3 & 1 \end{pmatrix}$ (ii) $\begin{pmatrix} 1 & 2 & 3 & 4 \\ 1 & 4 & 2 & 3 \end{pmatrix}$

(b) $\begin{pmatrix} 1 & 2 & 3 & 4 \\ 2 & 4 & 1 & 3 \end{pmatrix}$

(c) 4

(d) $X = \begin{pmatrix} 1 & 2 & 3 & 4 \\ 3 & 4 & 2 & 1 \end{pmatrix}$

4) (a) $\begin{pmatrix} 1 & 2 & 3 & 4 & 5 \\ 1 & 2 & 3 & 4 & 5 \end{pmatrix}$

(b) (i) order of group = order of generator (since group is cyclic) = 6

    (ii) $\begin{pmatrix} 1 & 2 & 3 & 4 & 5 \\ 3 & 1 & 2 & 4 & 5 \end{pmatrix}$

    (iii) $\begin{pmatrix} 1 & 2 & 3 & 4 & 5 \\ 3 & 1 & 2 & 5 & 4 \end{pmatrix}$

5) (a) order = lcm(3,5,2) = 30

(b) $\begin{pmatrix} 1 & 2 & 3 & 4 & 5 & 6 & 7 & 8 & 9 & 10 \\ 3 & 4 & 7 & 6 & 9 & 8 & 1 & 10 & 5 & 2 \end{pmatrix}^3$

$= \begin{pmatrix} 1 & 2 & 3 & 4 & 5 & 6 & 7 & 8 & 9 & 10 \\ 1 & 8 & 3 & 10 & 9 & 2 & 7 & 4 & 5 & 6 \end{pmatrix}$

$= (1)(2\ 8\ 4\ 10\ 6\ )(3)(7)(5\ 9)$

6) (a) (i) $\begin{pmatrix} 1 & 2 & 3 & 4 \\ 2 & 4 & 3 & 1 \end{pmatrix}^{-1} = \begin{pmatrix} 1 & 2 & 3 & 4 \\ 4 & 1 & 3 & 2 \end{pmatrix}$

(ii) $\begin{pmatrix} 1 & 2 & 3 & 4 \\ 2 & 4 & 3 & 1 \end{pmatrix}\begin{pmatrix} 1 & 2 & 3 & 4 \\ 3 & 4 & 1 & 2 \end{pmatrix} = \begin{pmatrix} 1 & 2 & 3 & 4 \\ 3 & 1 & 2 & 4 \end{pmatrix}$

$= (1\ 3\ 2)(4)$ or $(1\ 3\ 2)$

(b) (i) 3   (ii) 2

7) (i) (1 4 2 3)

(ii)

*	I	P	Q	R
I	I	P	Q	R
P	P	Q	R	I
Q	Q	R	I	P
R	R	I	P	Q

Table is closed.
Identity *I*.
I appears once in every row and once in every column;
therefore every element has a unique inverse.
Associativity is given.
Hence the table forms a group.

8) (a)

$+_6$	0	1	2	3	4	5
0	0	1	2	3	4	5
1	1	2	3	4	5	0
2	2	3	4	5	0	1
3	3	4	5	0	1	2
4	4	5	0	1	2	3
5	5	0	1	2	3	4

(b) Table is closed.
Identity = 0
0 appears once in every row and once in every column;
therefore every element has a unique inverse.
Addition is associative.
Hence the table forms a group.

(c) $5^1 = 5$
$5^2 = 4$
$5^3 = 3$
$5^4 = 2$
$5^5 = 1$
$5^6 = 0$

Order of element 5 = 6
   = order of group.
Hence cyclic.

Generators are 5 and 1

(d)

$+_6$	0	3
0	0	3
3	3	0

$+_6$	0
0	0

$+_6$	0	2	4
0	0	2	4
2	2	4	0
4	4	0	2

9) (a) xy, y^2

(b) $x^3 y^4 = x^2 x\, y^3 y$
$$= e\, x\, e\, y$$
$$= x\, y$$
$$= y\, x$$

(c) $(xy^2)^6 = x^6 y^{12}$ since * is commutative (the group is Abelian)
$$= (x^2)^3 (y^3)^4$$
$$= (e)^3 (e)^4$$
$$= e$$

Hence order is 6

(d) Order of element $xy^2 = 6$.
Order of group = 6.
Hence cyclic.

10) (a)

×₅	1	2	3	4
1	1	2	3	4
2	2	4	1	3
3	3	1	4	2
4	4	3	2	1

Table is closed
Identity = 1
1 appears once in every row and once in every column
therefore every element has a unique inverse.
Multiplication is associative.
Hence the table forms a group.

$$2^1 = 2$$
$$2^2 = 4$$
$$2^3 = 3$$
$$2^4 = 1$$

Order of element $2 = 4$.   Order of group $= 4$.   Hence cyclic.

(b) (i)

×₆	1	2	3	4	5
1	1	2	3	4	5
2	2	4	0	2	4
3	3	0	3	0	3
4	4	2	0	4	2
5	5	4	3	2	1

(ii) Cayley table is not closed. Hence it is not a group.

11) (a) Consider $(a*b^{-1})*(b*a^{-1})$

$$(a*b^{-1})*(b*a^{-1}) = a*(b^{-1}*b)*a^{-1} \quad \text{by associativity}$$
$$= a*e*a^{-1} \quad\quad\quad \text{definition of inverse}$$
$$= a*a^{-1} \quad\quad\quad\quad \text{definition of identity}$$
$$= e \quad\quad\quad\quad\quad\quad \text{definition of inverse}$$

Hence $[a*b^{-1}]^{-1} = b*a^{-1}$

(b) $aRa \Leftrightarrow a*a^{-1} \in H.$

   $\Rightarrow e \in H$ which is true since H must contain the identity.

Hence R is reflexive.

$aRb \Leftrightarrow a*b^{-1} \in H$

   $\Rightarrow [a*b^{-1}]^{-1} \in H$ since the inverse of any element in H
   must also be in H

   $\Rightarrow b*a^{-1} \in H$

Hence R is symmetric.

If $aRb \Leftrightarrow a*b^{-1} \in H$ and $bRc \Leftrightarrow b*c^{-1} \in H$

then $(a*b^{-1})*(b*c^{-1}) \in H$ since H is closed

   $\Rightarrow a*(b^{-1}*b)*c^{-1} \in H$

   $\Rightarrow a*e*c^{-1} \in H$

   $\Rightarrow a*c^{-1} \in H$

Hence R is transitive.

R reflexive, symmetric, transitive $\Rightarrow R$ is an equivalence relation.

12)   Let $(G,*)$ be a cyclic group with generator x

If x is a generator then so is x^{-1}

Since $(G,*)$ has only one generator : $x = x^{-1}$

Hence $x * x = x^{-1} * x$

$$= e$$

Therefore $(G,*)$ has only two elements $\{e, x\}$

13)   *Closure:* $\dfrac{1+2a}{1+2b} \times \dfrac{1+2c}{1+2d} = \dfrac{1+2a+2c+4ac}{1+2b+2d+4bd}$

$$= \dfrac{1+2(a+c+2ac)}{1+2(b+d+2bd)}$$

Since $a+c+2ac$ is an integer if a and c are integers, and $b+d+2bd$ is an integer if b and d are integers, we have an element of the correct form. Hence closure.

Identity :   Identity for multiplication $= 1$

Here $\dfrac{1+2x}{1+2x} = 1$

Inverse: Every element has an inverse $\dfrac{1+2y}{1+2x}$

since $\dfrac{1+2x}{1+2y} \times \dfrac{1+2y}{1+2x} = 1$ and $\dfrac{1+2y}{1+2x} \times \dfrac{1+2x}{1+2y} = 1$

Associativity: Multiplication is associative.

Hence group.

Exercise 6

1) (a) $cis\ 0$, $cis\dfrac{2\pi}{5}$, $cis\dfrac{4\pi}{5}$, $cis\dfrac{6\pi}{5}$, $cis\dfrac{8\pi}{5}$

(b)

$\times$	1	$cis\dfrac{2\pi}{5}$	$cis\dfrac{4\pi}{5}$	$cis\dfrac{6\pi}{5}$	$cis\dfrac{8\pi}{5}$
1	1	$cis\dfrac{2\pi}{5}$	$cis\dfrac{4\pi}{5}$	$cis\dfrac{6\pi}{5}$	$cis\dfrac{8\pi}{5}$
$cis\dfrac{2\pi}{5}$	$cis\dfrac{2\pi}{5}$	$cis\dfrac{4\pi}{5}$	$cis\dfrac{6\pi}{5}$	$cis\dfrac{8\pi}{5}$	1
$cis\dfrac{4\pi}{5}$	$cis\dfrac{4\pi}{5}$	$cis\dfrac{6\pi}{5}$	$cis\dfrac{8\pi}{5}$	1	$cis\dfrac{2\pi}{5}$
$cis\dfrac{6\pi}{5}$	$cis\dfrac{6\pi}{5}$	$cis\dfrac{8\pi}{5}$	1	$cis\dfrac{2\pi}{5}$	$cis\dfrac{4\pi}{5}$
$cis\dfrac{8\pi}{5}$	$cis\dfrac{8\pi}{5}$	1	$cis\dfrac{2\pi}{5}$	$cis\dfrac{4\pi}{5}$	$cis\dfrac{6\pi}{5}$

(c) $(\mathbb{Z}_5 , +)$ can be shown on the following Cayley table:

$+_5$	0	1	2	3	4
0	0	1	2	3	4
1	1	2	3	4	0
2	2	3	4	0	1
3	3	4	0	1	2
4	4	0	1	2	3

$$1 \leftrightarrow 0$$

$$cis\frac{2\pi}{5} \leftrightarrow 1$$

$$cis\frac{4\pi}{5} \leftrightarrow 2$$

$$cis\frac{6\pi}{5} \leftrightarrow 3$$

$$cis\frac{8\pi}{5} \leftrightarrow 4$$

Hence the two groups are isomorphic.

2) (i) Homomorphism requires $f(a \times b) = f(a) + f(b)$
for all $a, b \in \mathbb{R} - \{0\}$

$$\begin{aligned} f(a \times b) &= \log|ab| \\ &= \log|a| + \log|b| \\ &= f(a) + f(b) \end{aligned}$$

(ii) Identity for $(\mathbb{R}, +)$ is 0

Hence kernal of f is $\{-1, 1\}$

3) (a)

$+_4$	0	1	2	3
0	0	1	2	3
1	1	2	3	0
2	2	3	0	1
3	3	0	1	2

$\times_5$	1	2	3	4
1	1	2	3	4
2	2	4	1	3
3	3	1	4	2
4	4	3	2	1

(b) $0 \leftrightarrow 1$
        $2 \leftrightarrow 4$ (they are both self-inverses in their respective tables)
        $1 \leftrightarrow 3$
        $3 \leftrightarrow 2$

   or

        $0 \leftrightarrow 1$
        $2 \leftrightarrow 4$
        $1 \leftrightarrow 2$
        $3 \leftrightarrow 3$

4) $f(a+b) = 2^{a+b}$
   $\qquad\quad = 2^a \, 2^b$
   $\qquad\quad = f(a) \times f(b)$

   Hence homomorphism

   $f(a) = f(b)$ gives $2^a = 2^b$ $\Rightarrow a = b$

   Hence bijection

   Homomorphism + bijection = isomorphism

5)   (i) $f(a + b) = i^{a+b}$

$$= i^a\, i^b$$

$$= f(a) \times f(b) \qquad \text{Hence homomorphism}$$

     (ii)   Kernal $= \{\ldots\ldots\ -16,\ -8,\ -4,\ 0,\ 4,\ 8,\ 16,\ \ldots\ldots\ \}$

6) $f(a + b) = a + b + 1$

    $f(a) + f(b) = a + 1 + b + 1 = a + b + 2$

    Since $f(a + b) \neq f(a) + f(b)$ we do not have a homomorphism.

7) Homomorphism $\iff f(a * b) = f(a) * f(b)$

$$\iff (a*b)^{-1} = a^{-1} * b^{-1}$$

$$\iff (ab)^{-1} = a^{-1} b^{-1}$$

$$\iff ab\,(ab)^{-1} = ab\,a^{-1} b^{-1}$$

$$\iff e = ab\,a^{-1} b^{-1}$$

$$\iff eb = ab\,a^{-1} b^{-1} b$$

$$\iff b = ab\,a^{-1} e$$

$$\iff b = ab\,a^{-1}$$

$$\iff ba = ab\,a^{-1} a$$

$$\iff ba = abe$$

$$\iff ba = ab$$

$$\iff (G, *) \text{ is Abelian}$$

8 (a)

+$_6$	0	3
0	0	3
3	3	0

(b)

+$_6$	0	2	4
0	0	2	4
2	2	4	0
4	4	0	2

{0, 2, 4 } is a subset of {0, 1, 2, 3, 4, 5}

Table is closed; no new elements formed

Identity = 0

0 appears once in every row and once in every column;
therefore every element has an inverse.

Addition is associative.

Hence subgroup

(c) $0H = \{0 + 0,\ 0 + 2,\ 0 + 4\} = \{0, 2, 4\}$ (mod 6)

$1H = \{1 + 0,\ 1 + 2,\ 1 + 4\} = \{1, 3, 5\}$ (mod 6)

$2H = \{2 + 0,\ 2 + 2,\ 2 + 4\} = \{2, 4, 0\}$ (mod 6)

$3H = \{3 + 0,\ 3 + 2,\ 3 + 4\} = \{3, 5, 1\}$ (mod 6)

$4H = \{4 + 0,\ 4 + 2,\ 4 + 4\} = \{4, 0, 2\}$ (mod 6)

$5H = \{5 + 0,\ 5 + 2,\ 5 + 4\} = \{5, 1, 3\}$ (mod 6)

Hence left cosets are {0, 2, 4} and   {1, 3, 5}

www.ingramcontent.com/pod-product-compliance
Lightning Source LLC
Chambersburg PA
CBHW071818200526
45169CB00018B/409